Chemistry of High-Temperature Superconductors

CHEMISTRY OF HIGH-TEMPERATURE SUPERCONDUCTORS

Edited by:

Dr. Saeed Farrokhpay

ARCLER PRESS

www.arclerpress.com

Chemistry of High-Temperature Superconductors

Dr. Saeed Farrokhpay

Arcler Press
224 Shoreacres Road
Burlington, ON L7L 2H2
Canada
www.arclerpress.com
Email: orders@arclereducation.com

ISBN: 978-1-77469-823-5 (Hardcover)

ABOUT THE EDITOR

Dr Saeed Farrokhpay is a Chemical Engineer with several years of experience in mineral & material processing. He obtained his PhD from University of South Australia in 2005. He is currently a Technical Consultant in Australia. He has worked for more than 20 years at mineral and chemical industries, universities and research centers around the world. Dr Farrokhpay has published more than 90 papers in high ranked journals and conference proceedings. He has also edited several technical and scientific books, and served as an editorial board member of several international scientific journals.

TABLE OF CONTENTS

LIST OF FIGURES

LIST OF TABLES

PREFACE

High-temperature superconductors are a group of materials that defy conventional understanding of superconductivity. Unlike traditional superconductors, which require extremely low temperatures near absolute zero, high-temperature superconductors can achieve superconductivity at relatively higher temperatures, although still below room temperature. These materials typically consist of copper oxide or iron-based compounds, with complex crystal structures that give rise to their unique properties. The chemistry of high-temperature superconductors involves carefully tuning the composition, doping, and crystal structure to optimize the conditions for superconductivity. The interplay between electron-electron interactions, lattice vibrations, and magnetic properties is crucial in understanding the underlying mechanisms behind their superconducting behavior. Advances in the chemistry of high-temperature superconductors have the potential to revolutionize various fields, including energy transmission, magnet technology, and quantum computing. Ongoing research continues to explore novel materials and deepen our understanding of the intricate chemistry driving high-temperature superconductivity.

The book is structured into eight chapters, each focusing on a crucial aspect of high-temperature superconductivity:

Chapter 1 serves as an introduction to superconductors, providing a general overview of the field. It explains the fundamental concepts, historical context, and key properties associated with superconductivity, setting the stage for a deeper exploration of high-temperature superconductors. Chapter 2 delves into the general theory of high-TC (critical temperature) superconductors. It explores the theoretical frameworks and models used to understand the behavior of these materials at elevated temperatures, shedding light on the unique mechanisms that enable high-temperature superconductivity.

Chemistry plays a vital role in the understanding and development of high-temperature superconductors, and Chapter 3 focuses on the chemistry behind these remarkable materials. It investigates the composition, structure, and electronic properties of high-temperature superconductors, highlighting the interplay between chemical factors and superconducting behavior. Chapter 4 delves into the materials aspects of high-temperature superconductors for practical applications. It explores the engineering considerations, including the design, fabrication, and characterization of superconducting materials, with a specific focus on their potential applications in various fields.

The synthesis and fabrication techniques employed in the creation of high-temperature superconductors are the focus of Chapter 5. Chapter 6 explores the crystal structures of cuprate superconductors, which are a prominent class of high-temperature superconductors. It delves into the intricacies of their crystallography, highlighting the structural features that contribute to their superconducting properties.

Practical applications of high-temperature superconductors are covered in Chapter 7. Lastly, Chapter 8 explores emerging research and innovative directions in high-temperature superconductivity.

This book brings together a diverse group of experts in the field of high-temperature superconductivity. Their collective knowledge, research findings, and practical insights enrich each chapter, providing readers with a comprehensive understanding of the chemistry and applications of high-temperature superconductors.

CHAPTER

1 Introduction to Superconductors

Contents

1.1. INTRODUCTION

The physical phenomenon of superconductivity is a captivating occurrence that manifests in specific materials upon being subjected to cooling beneath a critical temperature. The hallmark feature of superconductors is their capacity to facilitate the flow of electrical current without experiencing any energy dissipation due to resistance. In contrast to conventional conductors, which consistently present a degree of resistance to the passage of electrical current, superconductors demonstrate a complete absence of electrical resistance (Hott, 2003).

Cooper pairing occurs when an electric current traverses a superconductor, leading to the formation of electron pairs responsible for the conduction of the current. The collective movement of these pairs within the lattice of the superconducting substance facilitates the efficient conduction of electrical charge, thereby avoiding the dispersion and collisions commonly responsible for resistance in conventional conductors. Superconductors are known for their high efficiency in conducting electricity due to the absence of heat dissipation, resulting in zero energy loss. The occurrence of zero electrical resistance in superconductors holds noteworthy implications. The technology enables the development of superconductive circuits that exhibit negligible power losses during electricity transmission over extended distances. The aforementioned characteristics also facilitates the production of robust magnetic fields by implementing superconducting coils. These fields have practical uses in domains such as particle accelerators and magnetic resonance imaging (MRI) (Paul et al., 2001) (Figure 1.1).

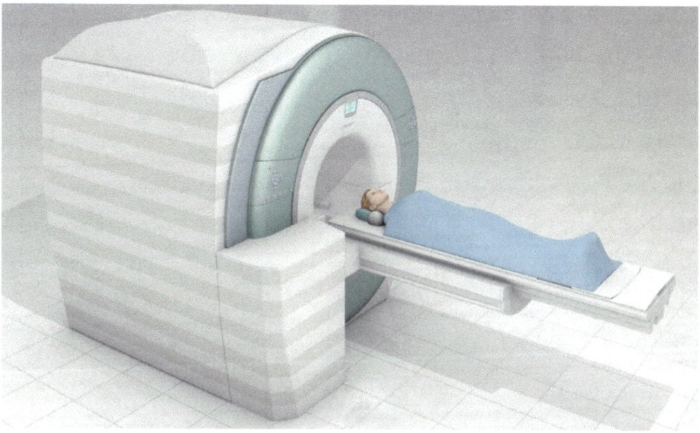

Figure 1.1. *Image of magnetic resonance imaging (MRI) scan.*

Source: Hopkins medicine, Creative Commons License.

The critical temperature, commonly called Tc, is a fundamental characteristics of superconductors. The term denotes the critical temperature at which a substance undergoes a phase transition into the state of superconductivity, characterized by the absence of electrical resistance. Upon surpassing this crucial temperature, the substance exhibits typical conductive behavior with measurable resistance. The critical temperature exhibits significant variability across diverse materials.

The initial superconductors that were identified during the early 1900s, such as lead and mercury, exhibited critical temperatures that were notably low (Saxena, 2012). As a result, it was necessary to cool these materials to temperatures close to absolute zero (–273.15°C) to observe the phenomenon of superconductivity.

The field underwent a significant transformation with the emergence of superconductors with high temperatures in the late 1980s. The critical temperatures of high-temperature superconductors are notably greater than those of conventional superconductors. The aforementioned materials can manifest superconductivity at temperatures surpassing the boiling point of liquid nitrogen (–196°C), rendering them significantly more feasible and pragmatic for many applications (Wesche, 2015).

Tip

Understanding the fundamental principles of superconductivity is crucial for exploring its wide range of applications.

The capacity to attain superconductivity at elevated temperatures has broadened the scope for the extensive integration and utilization of superconductors in various domains, including energy distribution, medical diagnostics, and quantum computation. The threshold temperature of a superconductor is subject to various influences, such as the constituent material, crystal lattice configuration, and the existence of extrinsic magnetic fields.

Scholars investigate and manipulate these variables to explore and enhance superconducting substances with elevated critical temperatures. Through the progression of our comprehension of superconductivity and the creation of materials with elevated critical temperatures, it is possible to unveil further pragmatic and effective implementations for these extraordinary substances (Krabbes, Fuchs, Canders, May, & Palka, 2006).

1.2. HISTORICAL BACKGROUND

1.2.1. Discovery of Superconductivity

In the year 1911, the Dutch physicist Heike Kamerlingh Onnes achieved a noteworthy advancement in comprehending the characteristics of materials under conditions of exceedingly low temperatures. Onnes conducted experiments at the University of Leiden to look into the electrical resistance of different substances as they came upon absolute zero (Qiu, 2011). The researcher lowered the temperature of mercury to levels in proximity to absolute zero and documented a sudden and notable decrease in its electrical resistance. The first observation of superconductivity was marked by the occurrence of an unforeseen phenomenon in which the resistance dissipated entirely. The term "superconductivity" was introduced by Onnes to characterize the distinctive phenomenon exhibited by specific materials under conditions of low temperature (Mitschang, 1995).

1.2.2. Key Milestones in Superconductivity Research

Throughout the years, the advancement of superconductivity research has been influenced by various significant milestones:

1.2.2.1. 1933: Meissner-Ochsenfeld Effect

The year 1933 marked a notable advancement in comprehending superconductors, as German physicists Fritz and Heinz London made a noteworthy contribution to this field. The notion of perfect diamagnetism was introduced, elucidating the phenomenon of magnetic field exclusion from the within of superconducting substances. The Meissner effect, which was observed, illustrates the capacity of superconductors to exclude magnetic fields from their interior completely (ŞAKİROĞLU, 2002) (Figure 1.2).

Remember

Superconductors exhibit zero electrical resistance, enabling the efficient flow of electric current.

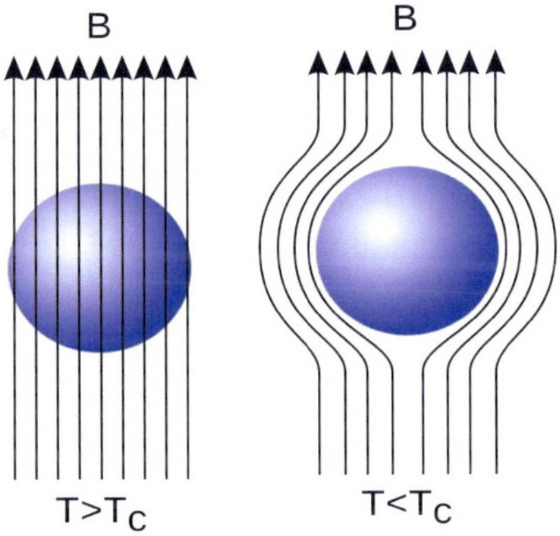

Figure 1.2. *Illustration of Meissner effect.*

Source: Piotr Jaworski, Creative Commons License.

1.2.2.2. 1950: BCS Theory

The Bardeen-Cooper-Schrieffer (BCS) theory of superconductivity was formulated by a trio of American physicists, namely John Bardeen, Leon Cooper, and John Robert Schrieffer, in the year 1950. This innovative theory offers a comprehensive elucidation for the conduct of superconducting materials. According to the BCS theory, the phenomenon of superconductivity can be attributed to the pairing of electrons, which is believed to occur due to their interaction with lattice vibrations, commonly referred to as phonons (Balian, Flocard, & Vénéroni, 1999). The aforementioned text effectively explicated diverse phenomena that are commonly observed in superconductors, such as the lack of electrical resistance and the Meissner effect. The Nobel Prize in Physics was awarded to Bardeen, Cooper, and Schrieffer in 1972 for their collective contributions to the field (Hirsch, 2009).

1.2.2.3. 1986: High-Temperature Superconductors

In 1986, a noteworthy breakthrough was achieved in the field of superconductivity research by Georg Bednorz and Alex Müller, who were affiliated with IBM at the time. They identified a novel category of substances that exhibited superconductivity at high temperatures. The discovery made by Bednorz and Müller pertains to the superconductivity of a ceramic material based on copper, which was observed to occur at temperatures surpassing the boiling point of liquid nitrogen, specifically at 77°K or –196°C (Müller & Bednorz, 1987). Superconductivity was predominantly observed at temperatures in close proximity to absolute zero prior to the aforementioned discovery. The identification of high-temperature superconductors has presented novel prospects for pragmatic implementations and has engendered extensive enthusiasm and investigation within the discipline (Lancaster, 1997) (Figure 1.3).

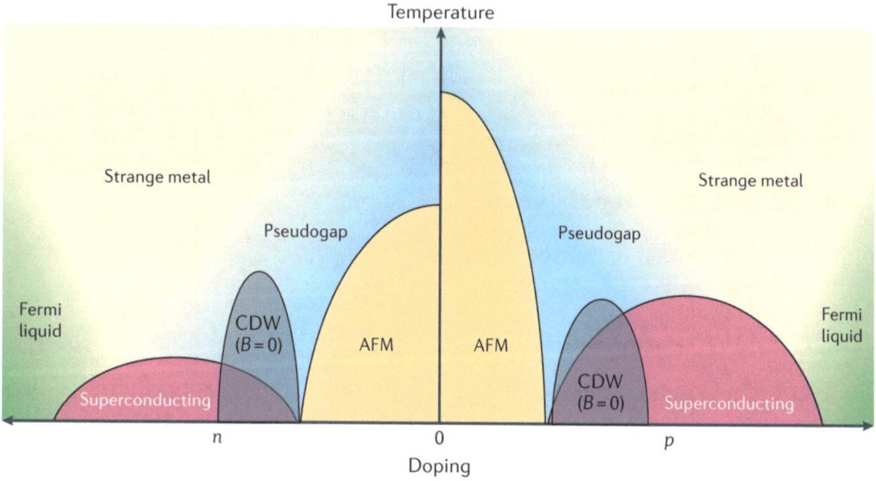

Figure 1.3. *Schematic of high-temperature superconductivity.*

Source: Xingjiang, Creative Commons License.

1.2.3. Prominent Scientists and Their Contributions

The notable contributions of several scientists have enriched the discipline of superconductivity. The following are noteworthy persons and their respective contributions.

1.2.3.1. Heike Kamerlingh Onnes

The finding of superconductivity is attributed to Onnes, a physicist from the Netherlands. The individual's innovative investigations involving mercury resulted in recognition of the phenomenon, ultimately leading to the individual being awarded the Nobel Prize in Physics in 1913. The contributions of Onnes established the fundamental basis for subsequent investigations and inquiries into the phenomenon of superconductivity (Onnes, Leiden, & Onnes, 2016).

1.2.3.2. Fritz and Heinz London

Fritz and Heinz London have significantly contributed to the comprehension of superconductivity. The notion of perfect diamagnetism was first introduced in 1933 to elucidate the Meissner effect observed in superconductors. The research conducted by the individuals in question yielded significant revelations regarding the electromagnetic characteristics of superconductors and developed basic concepts that remain relevant in contemporary times (London & London, 1935).

1.2.3.3. John Bardeen, Leon Cooper, and John Robert Schrieffer

The BCS theory of superconductivity, formulated by Bardeen, Cooper, and Schrieffer, has significantly transformed the domain of superconductivity. The theory, which was developed in 1950, elucidated the phenomenon of superconductivity by attributing it to the pairing of electrons that resulted from their relationship with lattice vibrations, commonly referred to as phonons. The BCS theory, which garnered the Nobel Prize in Physics in 1972, furnished a comprehensive framework for comprehending superconductivity (Bardeen, Cooper, & Schrieffer, 1957).

1.2.3.4. Georg Bednorz and Alex Müller

The identification of high-temperature superconductors by Bednorz and Müller in 1986 represented a significant breakthrough in the field of superconductivity investigation. The discovery made by the researchers has paved the way for attaining superconductivity at elevated temperatures that are more feasible. The aforementioned revelation instigated a notable upswing in scholarly investigation and examination of high-temperature superconductors, opening avenues for prospective utilization in diverse domains (Georg Bednorz & Müller, 1988).

The aforementioned researchers, among numerous others, have made noteworthy advancements in the realm of superconductivity. The research conducted by the authors has contributed significantly to the advancement of our comprehension of the phenomenon, provided direction for subsequent investigations, and facilitated the progress of superconducting materials and technological advances. The impact of their contributions persists as a source of inspiration and influence for contemporary advancements in the field (J George Bednorz & Müller, 1986).

1.3. FEATURES AND PROPERTIES OF SUPERCONDUCTORS

Superconductors exhibit exceptional features and attributes that set them apart from conventional substances. Comprehending the properties in question is of utmost importance in gaining a comprehensive understanding of the fundamental nature of superconductivity and its practical applications. This section will explore three fundamental characteristics of superconductors. The phenomenon of zero electrical resistance and ongoing currents, ideal diamagnetism and the Meissner effect, as well as the vital magnetic field and critical current density, are noteworthy characteristics in the field of physics (Anlage, 2010).

1.3.1. Zero Electrical Resistance and Persistent Currents

Superconductors are characterized by their remarkable property of displaying zero electrical resistance, which allows for the unimpeded conduction of electric current. The distinctive phenomenon under consideration emerges due to the creation of Cooper pairs. These pairs consist of two electrons with opposite spins and momenta, which are held together by the attractive forces facilitated by lattice vibrations, also known as phonons. During the state of superconductivity, the Cooper pairs exhibit coherent motion throughout the material, without experiencing any scattering due to contaminants or lattice vibrations. Consequently, the phenomenon of electrical resistance reduction to nil enables the smooth and effective conveyance of electric current (Halperin, Refael, & Demler, 2010).

Moreover, superconductors exhibit the capability to maintain continuous electric currents. Upon application of an electric current to a superconducting loop, the current can persist indefinitely without necessitating any external power supply. The continuous flow of current results from the lack of electrical resistance, which precludes the release of energy in the form of thermal energy. Persistent currents find practical applications in various fields, such as magnetic energy storage systems, superconducting magnets employed in particle accelerators, and magnetic resonance imaging (MRI) machines (Little, 1967) (Figure 1.4).

Did You Know?

High-temperature superconductivity refers to the discovery of materials that can superconduct at temperatures significantly higher than traditional superconductors.

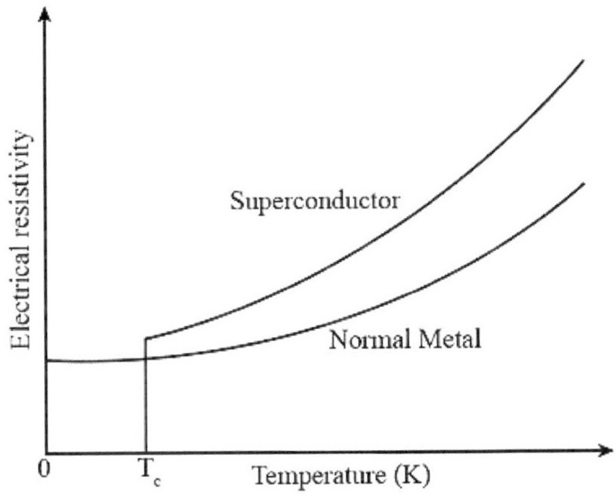

Figure 1.4. *Graphical representation of zero electrical resistance associated with high-temperature superconductors.*

Source: Homework. Study, Creative Commons License.

1.3.2. Perfect Diamagnetism and the Meissner Effect

Superconductors display a distinctive characteristics known as perfect diamagnetism upon being cooled beneath their critical temperature. Diamagnetism pertains to the capacity of a substance to generate a counteractive magnetic field in response to an externally applied magnetic field. The diamagnetic response exhibited by superconductors is notably robust, leading to the ejection of magnetic flux from the material's interior (Essén & Fiolhais, 2012). The phenomenon referred to as the Meissner effect is a consequence of the emergence of Cooper pairs and their reaction to magnetic fields. When a superconductor is exposed to an external magnetic field, the Meissner effect induces a complete expulsion of the magnetic field from the material, leading to shielding of the interior from any external magnetic fields. The superconductor displays robust diamagnetic characteristics, whereby the magnetic field lines curve around the substance as a result. The Meissner effect is a phenomenon that finds practical applications in various technologies, including levitation of magnetic objects and the development of potent, high-field magnets (Hirsch, 2016).

1.3.3. Critical Magnetic Field and Critical Current Density

Superconductors are subject to limitations with respect to their capacity to maintain their superconducting characteristics while carrying a current and sustaining a magnetic field. The critical magnetic field (Hc) denotes the upper limit of magnetic field strength beyond which a superconductor transitions from a state of superconductivity to a state of normal conductivity. The determination of Hc is contingent upon the material's inherent properties, which are subject to fluctuations based on variables such as its composition, crystal lattice configuration, and thermal conditions. The critical current density (Jc) is defined as the upper limit of current density that a superconductor can sustain without transitioning to a non-superconducting state (van der Laan et al., 2010). The behavior of Jc is contingent upon a multitude of factors, such as temperature, material characteristics, and the existence of flaws or contaminants. The critical parameters of Hc and Jc hold significant importance in both the design and functioning of superconducting devices. Increased amounts of the critical magnetic field and critical current density are advantageous as they facilitate the advancement of more robust magnets and enhanced electrical systems. The progress made in developing superconductor materials and their engineering has resulted in noteworthy enhancements in these crucial parameters. This has broadened the scope of possible applications in various fields, including power transmission, magnetic resonance imaging (MRI), and superconducting electronics (Heine, Tenbrink, & Thöner, 1989) (Figure 1.5).

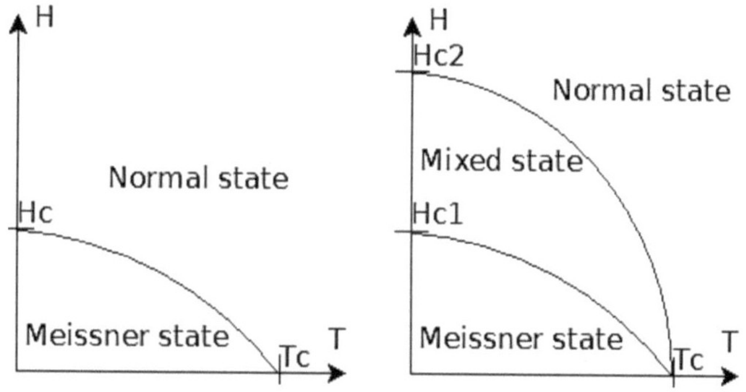

Figure 1.5. *Graphical representation of critical magnetic field.*

Source: Łukasz Tomków, Creative Commons License.

To summarize, superconductors exhibit exceptional characteristics that distinguish them from conventional substances. The characteristics of the zero electrical resistance property and the capacity to maintain persistent currents, the perfect diamagnetism displayed through the Meissner effect, and the critical parameters encompassing the magnetic field and current density limits. Comprehending and controlling these characteristics are imperative in exploiting the complete capabilities of superconductors in diverse pragmatic implementations throughout various scientific and technological domains (Bugoslavsky et al., 2001).

1.4. CATEGORIES OF SUPERCONDUCTORS

Superconductors can be categorized into two distinct types, namely Type I and Type II superconductors. The categorizations are established on the distinct behaviors and traits displayed by these substances during the occurrence of the superconducting phase shift.

1.4.1. Type I Superconductors

The process of moving to the superconducting state in Type I superconductors is distinguished by a discernible behavior. Following a critical temperature (Tc), they demonstrate an abrupt and total cessation of electrical resistance. Furthermore, Type I superconductors exhibit the Meissner effect, which is the expulsion of magnetic fields from their interior (Mele et al., 2019).

1.4.1.1. Characteristics and Behavior

Type I superconductors exhibit a number of noteworthy features. Primarily, superconductors exhibit a distinct and abrupt shift from their regular state to the superconducting state, which transpires at a clearly defined critical temperature. At temperatures below the critical point, the electrical resistance exhibits a significant decrease to the point of reaching zero, thereby facilitating the unimpeded conduction of electric current. Nevertheless, beyond the critical temperature, they exhibit typical conductive behavior characterized by a measurable resistance (Prozorov, Fidler, Hoberg, & Canfield, 2008).

The attribute that distinguishes Type I superconductors are their flawless diamagnetic behavior. Type I superconductors display a complete displacement of the magnetic field from their interior when exposed to a magnetic field from the outside below their critical temperature. The phenomenon of expulsion is attributed to the emergence of a screening current at the material's surface, resulting in the generation of a magnetic field that opposes the external field and ultimately nullifies it (Strongin, Paskin, Schweitzer, Kammerer, & Craig, 1964).

1.4.1.2. Examples and Materials

Type I superconductors comprise of elemental metals, including mercury

(Hg), lead (Pb), and tin (Sn), among others. The aforementioned metals demonstrate Type I characteristics upon cooling below their critical temperatures. Furthermore, Type I superconductors encompass specific binary and ternary alloys, such as niobium-tin (Nb_3Sn) and niobium-titanium (NbTi) (Babaev & Speight, 2005).

1.4.2. Type II Superconductors

Type II superconductors display a more intricate set of characteristics when compared with Type I superconductors. Superconductors with a broader range of stability in their superconducting state and the ability to withstand higher magnetic fields without compromising their superconducting properties are available. Type II superconductors can preserve a certain amount of magnetic flux inside their composition in the configuration of quantized vortices (Goodman, 1966) (Figure 1.6).

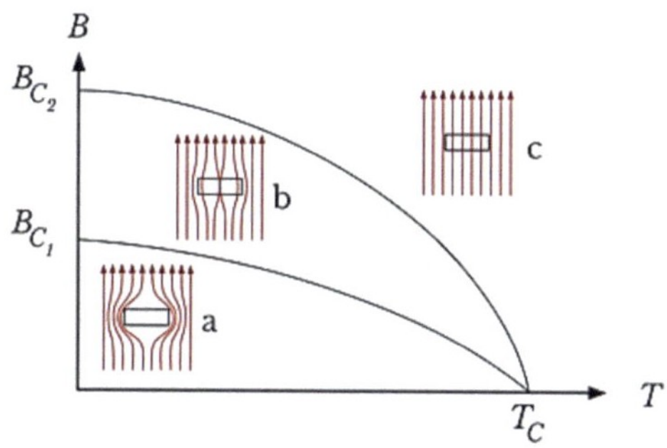

Figure 1.6. *Graphical representation of Type II superconductor.*

Source: Frederic Bouquet & Julien Bobroff, Creative Commons License.

1.4.2.1. Characteristics and Behavior

Type II superconductors gradually transition from their normal state to the superconducting state, which is distinguished by a wide temperature range in which superconductivity is observed. Superconductors demonstrate zero resistance when their temperature falls below a critical value (Tc). However, they can endure a certain degree of resistivity when subjected to strong magnetic fields (Rosenstein & Li, 2010).

In contrast to Type I superconductors, Type II superconductors do not exhibit complete expulsion of magnetic fields. Rather than prohibiting the penetration of magnetic flux, they permit its ingress into their interior by means of quantized vortices. The formation of a lattice structure by the vortices is accompanied by a quantified magnetic flux carried by each vortex. The capacity of Type II superconductors to sustain a certain amount of magnetic flux proves advantageous in diverse practical implementations (Sonier, Brewer, & Kiefl, 2000).

1.4.2.2. Examples and Materials

Type II superconductors encompass a variety of materials. Niobium-based compounds, specifically niobium-tin (Nb_3Sn) and niobium-titanium (NbTi), are noteworthy examples. The utilization of these compounds is prevalent in the fabrication of superconducting magnets owing to their elevated magnetic fields and temperatures.

Additional Type II superconductors include specific ceramic compounds recognized as high-temperature superconductors (HTS). Such materials include Yttrium-Barium-Copper-Oxide (YBCO) and Bismuth-Strontium-Calcium-Copper-Oxide (BSCCO). High-temperature superconducting (HTS) materials are distinguished by their ability to demonstrate superconductivity at temperatures that are comparatively higher than those of conventional low-temperature superconductors. This characteristics has the potential to create novel opportunities for practical applications (Dew-Hughes, 1974).

To summarize, Type I superconductors manifest a distinct shift to the superconducting state, and full repulsion of magnetic fields, and are frequently encountered in elemental metals and alloys. In contrast to Type I superconductors, Type II superconductors exhibit a wider range of transitions and permit the partial entry of magnetic fields in the form of vortices (Gropp et al., 1996). Examples of Type II superconductors include niobium-based compounds and high-temperature superconductors. The unique attributes and behaviors exhibited by these various superconductors render them well-suited for diverse applications and facilitate further progress in superconductivity (Rosenstein & Li, 2010).

1.5. BCS THEORY OF SUPERCONDUCTIVITY

The BCS theory, which was developed by John Bardeen, Leon Cooper, and John Robert Schrieffer in 1957, had a transformative impact on the comprehension of superconductivity. The aforementioned theory thoroughly elucidates the underlying mechanisms that give rise to superconductivity in traditional superconductors under conditions of reduced temperature. The fundamental concept underlying the BCS theory is the formation of Cooper pairs. The aforementioned pairs comprise a duo of electrons possessing opposing spins and momenta, which are conjoined through an alluring interaction (Hirsch, 2009).

The process of Cooper pair formation is made easier by the relationship among electrons and lattice vibrations, commonly referred to as phonons. As the electron traverses the lattice, it generates minor distortions in the lattice configuration, resulting in regional energy potential. The aforementioned potential serves as a force of attraction, facilitating the movement of an electron possessing opposing spin and momentum towards it. The phenomenon in question gives rise to the creation of a Cooper pair, wherein two electrons exhibit an attractive force resulting in a shared momentum of zero. The electron-phonon interaction significantly influences the process, which facilitates the attractive interaction between electrons (Johnston, 2013).

The energy gap constitutes an essential component of the BCS theory and assumes a crucial function in the phenomenon of superconductivity. The emergence of an energy gap in the electronic band structure of a superconductor is attributed to the creation of Cooper pairs. Within the superconducting state, an array of energies that electrons cannot occupy is commonly referred to as the energy gap. The aforementioned discrepancy emerges as a consequence of the singular coupling of electrons, leading to a discernible density of electronic states (Suhl, Matthias, & Walker, 1959).

The energy gap holds considerable significance in determining the conductive properties of electrons within a superconductive material. Superconductivity is observed below the critical temperature (Tc), at which point the energy gap effectively inhibits the excitation of individual electrons. Electron pairs possessing energies surpassing the energy gap are prohibited, whereas those that follow the energy gap are permissible. The imposition of this constraint hinders the interaction of electrons with contaminants or lattice defects, thereby resulting in the absence of electrical resistance. As a result, superconductors demonstrate ideal conductivity and possess the ability to transport electrical currents without any dissipation (Allen & Mitrović, 1983).

In addition, the energy gap plays a role in the manifestation of the Meissner effect, a distinctive superconductivity attribute. In the presence of a magnetic field below a critical threshold, the energy gap characteristics of superconductors impedes the creation of excited electron states that might possibly absorb the energy of the magnetic field. Consequently, the magnetic field is repelled from the superconductor's internal region, resulting in complete diamagnetism. The phenomenon of expelling magnetic fields, known as the Meissner effect, is accountable for the distinctive conduct of superconductors when subjected to external magnetic fields (Schrieffer, 2018).

The determination of the energy gap magnitude is contingent upon the potency of the electron-phonon relationship and the superconductor's critical temperature. Typically, heightened electron pairing increases energy gaps associated with elevated critical temperatures. Superconductors that function at elevated temperatures demonstrate greater energy gaps than traditional superconductors that operate at lower temperatures (Bertrand, 2005).

Comprehending the BCS theory and the import of the energy gap has played a pivotal role in propelling the domain of superconductivity forward. The aforementioned has established a basis for subsequent investigations, resulting in the identification and advancement of diverse superconductive substances possessing distinctive characteristics and utilities. The BCS theory is a fundamental concept in the investigation of superconductivity, enabling progress in various domains, including energy distribution, medical diagnostics, and quantum computation (Schmalian, 2010).

1.6. HIGH-TEMPERATURE SUPERCONDUCTIVITY

1.6.1. Challenges and Findings in High-Temperature Superconductivity

The term HTS) pertains to the occurrence of superconductivity at relatively elevated temperatures, typically surpassing the boiling point of liquid nitrogen (–196°C or 77°K). The aforementioned breakthrough has brought about a significant transformation in the realm of superconductivity, as it has enabled the utilization of this phenomenon in pragmatic scenarios at comparatively more feasible temperatures. Nevertheless, comprehending high-temperature superconductivity presents numerous obstacles (Orenstein & Millis, 2000).

Unravelling the fundamental mechanism responsible for high-throughput screening (HS) represents a significant challenge. The BCS theory elucidates the phenomenon of conventional superconductivity, which is predicated on creating Cooper pairs via electron-phonon couplings at extremely low temperatures. In contradistinction, the HTS phenomenon challenges the established comprehension, given that the temperatures of utmost importance are markedly greater than the values anticipated by the Bardeen-Cooper-Schrieffer (BCS) theory (Burns, 1992).

The comprehension and amalgamation of intricate crystal structures of high-temperature superconductors present an additional obstacle. Materials with HTS frequently display complex lattice structures that may give rise to conflicting interactions, impeding the development of superconducting states. Understanding the intricate relationship between structural properties and superconductivity is paramount in driving progress in this domain (Ginzburg, Kirzhnits, Agyei, & Birman, 1982).

Scientists have made noteworthy discoveries in the pursuit of achieving high-temperature superconductivity. In 1986, a significant advancement was made in the field of superconductivity by Georg Bednorz and Alex Müller, who identified the occurrence of superconductivity in a copper oxide compound, especially a lanthanum-based cuprate. The aforementioned finding served as the inception of the HS era and sparked widespread research endeavors on a global scale (Emery & Reiter, 1988).

Subsequently, many high-temperature superconductor families have been identified, each manifesting distinct characteristics and critical temperatures. Several noteworthy families of superconductors comprise the cuprates, iron-based superconductors, and diverse heavy fermion compounds. The aforementioned findings have expanded the comprehension of superconductivity and unveiled fresh prospects for technological implementations (Zhou et al., 2021).

1.6.2. Different Concepts and Mechanisms Proposed

Numerous hypotheses and processes have been posited to elucidate the phenomenon of superconductivity at elevated temperatures. Although a complete comprehension remains elusive, these theories offer helpful insights into the fundamental physics of high-temperature superconductivity.

The d-wave pairing theory is a prominent theoretical framework that is applicable to a multitude of cuprate superconductors. The proposal posits that the electron pairs that give rise to superconductivity exhibit an atypical "d-wave" symmetry, as opposed to the typical s-wave symmetry that is evident in superconductors operating at low temperatures. The aforementioned theory elucidates certain distinctive characteristics of cuprate superconductors, including the anisotropic nature of the superconducting gap (Lozovik & Yudson, 1976).

The strong electron-electron correlation theory has garnered considerable attention as a plausible explanation. The proposition posits that electron-electron interactions are pivotal in specific HTS materials, resulting in the development of superconductivity. The present theory elucidates the anomalous conduct exhibited by cuprates and heavy fermion compounds, wherein the Coulombic repulsion among electrons is remarkably robust.

Moreover, spin fluctuations have been posited as a potential mechanism for high-temperature superconductivity. The proposition is that superconductivity mediation relies upon interactions between magnets and spin fluctuations. The aforementioned theory is corroborated by the iron-based superconductors, wherein it is postulated that the proximity to magnetic order augments the superconducting characteristics (Chubukov, 2012).

It is imperative to acknowledge that the aforementioned theories do not possess a mutually exclusive nature, and the high-temperature superconductivity mechanism may potentially encompass a confluence of factors. Continual research and experimental inquiries persist in refining and broadening our comprehension of these intricate substances (Clarke, 1968).

1.6.3. Outline of Notable High-Temperature Superconductors

1.6.3.1. Cuprates

Cuprate superconductors, commonly called high-temperature superconducting copper oxides, have been the subject of extensive research since their initial discovery. $La_{2-x}Sr_xCuO_4$ (LSCO) is the initial high-temperature superconductor, exhibiting a critical temperature (Tc) of approximately 35 K. Additional cuprate families of significance comprise $YBa_2Cu_3O_7$ (YBCO) exhibiting a critical temperature (Tc) exceeding 90 K and $Bi_2Sr_2CaCu_2O_8$ (BSCCO) exhibiting a Tc exceeding 100 K. The cuprates demonstrate stratified configurations and intricate electronic characteristics, rendering them well-suited for investigating the underlying causes of high-temperature superconductivity (Kordyuk, 2015).

1.6.3.2. Iron-Based Superconductors

The iron-based superconductors represent another category of high-temperature superconductors. The aforementioned materials are composed of a number of layers of iron atoms that exhibit interactions with various elements, resulting in the emergence of superconductivity. The critical temperatures exhibited by iron-based superconductors are compound-specific and can vary between 20 K and above 50 K. Illustrative instances comprise of the 122 family, such as $BaFe_2As_2$, and the 1111 family, such as $LaFeAsO_{1-x}F_x$. The structural similarities of iron-based superconductors to other associated electron systems and the potential for unconventional pairing mechanisms have garnered considerable interest (Ren & Zhao, 2009).

1.6.3.3. Heavy Fermion Compounds

The heavy fermion superconductors are a distinctive category of high-temperature superconducting materials. The aforementioned compounds comprise elements of the rare earth or actinide series and display amplified electronic mass due to robust electron correlations. The critical temperatures exhibited by heavy fermion superconductors exhibit a broad range, spanning from a few millikelvin to in excess of 20 K. Notable instances comprise $CeCu_2Si_2$ and UPt_3. The significance of these materials lies in their amalgamation of features from the domains of superconductivity and strongly correlated electron physics (Petrovic et al., 2001).

In addition to these families, research is ongoing to identify and comprehend additional high-temperature superconductors, including organic superconductors, turned bilayer graphene, and hybrid systems.

1.7. LOW-TEMPERATURE SUPERCONDUCTORS (LTS)

1.7.1. Common Materials and Their Properties

Low-temperature superconductors (LTS) refer to a class of materials that demonstrate the phenomenon of superconductivity at comparatively low temperatures, usually falling beneath 30°K (–243.2°C) (Shein & Ivanovskii, 2009). The aforementioned materials have been instrumental in advancing superconductivity and are still being utilized in diverse domains. Several materials are commonly used in Long-Term Support applications:

1.7.1.1. Lead (Pb)

Lead is considered to be one of the earliest identified superconductors and holds the distinction of being among the primary materials in which the phenomenon of superconductivity was initially detected. The substance exhibits exceptional electrical and thermal conductivity characteristics and has a critical temperature (Tc) of approximately 7.2°K. Due to their low critical temperature (Tc), lead-based superconductors are predominantly utilised in low-temperature applications (Smith, Chu, & Maple, 1969).

1.7.1.2. Mercury (Hg)

Superconductivity in Mercury is observed at temperatures that are lower than 4.2°K. Frequently, it is combined with additional elements, like thallium or barium, to augment its superconductive characteristics. Superconductors based on mercury are utilized in diverse applications, such as cryogenics and fundamental research (Pippard, 1947).

1.7.1.3. Niobium-Titanium (NbTi)

The material known as NbTi is extensively utilized as a LTS, especially in the fabrication of superconducting magnets. With a Tc of approximately 9°K and remarkable mechanical properties, this material is well-suited for utilization in scenarios that demand robust and consistent magnetic fields. The composite material commonly utilized is NbTi, wherein superconducting filaments are incorporated into a copper matrix to augment its mechanical robustness (Larbalestier & Lee, 1995).

1.7.1.4. Niobium-Tin (Nb_3Sn)

Nb_3Sn is a notable low-temperature superconductor material that exhibits a critical temperature that is comparatively higher, typically within the range of 15 to 18°K. This material exhibits outstanding superconducting properties and demonstrates superior tolerance to magnetic fields when compared to NbTi. The compound Nb_3Sn finds utility in the fabrication of high-field magnets employed in diverse areas such as particle accelerators and fusion research (Lee & Larbalestier, 2003).

1.7.1.5. Other LS Materials

Several other Long-Term Storage (LS) materials are available, such as aluminum (Al), tin (Sn), and certain intermetallic compounds. Every substance possesses a distinct range of characteristics and is selected according to particular demands of the given application (Dagotto, 1994).

1.7.2. Fabrication Practices for LTS Materials

The production of Low Temperature Superconductor (LTS) materials entails a multitude of procedures that strive to generate a uniform, uncontaminated, and robust material exhibiting exceptional superconducting capabilities. Several fabrication techniques are commonly utilized for Low Temperature Superconducting (LTS) materials.

1.7.2.1. Powder Metallurgy

The technique of powder metallurgy is extensively employed in the production of low-temperature superconducting (LTS) materials. The process entails the generation of Long-Term Stable powders utilizing diverse techniques, including mechanical alloying and chemical synthesis. Subsequently, the aforementioned powders undergo compaction and sintering procedures at heightened temperatures, resulting in the formation of solid materials with regulated microstructures and high density. The process of powder metallurgy enables the fabrication of low-temperature superconductive (LTS) materials that exhibit favorable characteristics, including refined grain size and improved superconducting capabilities (Thieme, Pourrahimi, Schwartz, & Foner, 1984).

1.7.2.2. Wire Drawing

Superconducting wires are frequently utilized in the form of LTS materials. The process of wire drawing entails the extrusion of low-tensile-strength (LTST) materials through a sequence of dies, resulting in the reduction of their diameter and the creation of unbroken filaments. The process of wire drawing facilitates the alignment of grains in the material, thereby enhancing its superconducting characteristics. The wires that are obtained can undergo additional processing procedures to augment their mechanical durability, such as by integrating them within a copper matrix or applying protective coatings onto their surfaces (Lee & Larbalestier, 2003).

1.7.2.3. Thin Film Deposition

Thin films composed of Low Temperature Superconducting (LTS) materials are utilized in various applications, such as superconducting electronics and sensors. The process of depositing low-temperature superconductor (LTS) materials onto substrates with precise thickness and composition is achieved through the utilization of thin film deposition methods, including physical vapor deposition (PVD) and chemical vapor deposition (CVD). Utilizing these methodologies enables the production of slender layers possessing homogeneous characteristics and meticulous regulation of the film's arrangement and constitution (Jackson & Palmer, 1994).

1.7.2.4. In Situ Techniques

In certain instances, LTS materials are produced through in situ methodologies, whereby the fabrication process induces superconductivity. In the case of Nb_3Sn, a prevalent methodology entails subjecting a composite material comprising niobium and tin filaments to a heat treatment process. Heat application induces a chemical reaction between niobium and tin, leading to the emergence of the Nb_3Sn superconducting phase (Ma, 2012).

1.7.2.5. Optimization and Annealing

Following the primary manufacturing procedures, low-temperature superconducting (LTS) materials frequently undergo refinement techniques to enhance their superconductive characteristics. The process may entail annealing, a technique that involves the deliberate application of heat and subsequent cooling of the material in a controlled manner to improve the structure of its crystals and remove any existing defects. The annealing process has been observed to enhance the critical current density and diminish residual resistivity, thereby resulting in enhanced superconducting performance (Yamaguchi, Hata, Mochiku, & Yasuoka, 2012).

Producing LTS materials generally entails a fusion of methodologies, including powder metallurgy, wire drawing, thin film accumulation, and in situ procedures. Utilizing these methodologies facilitates the generation of low-temperature superconducting (LTS) materials possessing customized characteristics appropriate for diverse applications in the domains of superconducting components, magnets, and other fields reliant on low-temperature superconductivity (Chancellor, Zohren, & Warburton, 2017).

1.8. USES OF SUPERCONDUCTORS

1.8.1. Superconducting Electronics

The field of superconducting electronics investigates the potential implementation of superconducting materials and devices in diverse electronic applications. The distinctive characteristics exhibited by superconductors, namely their capacity for carrying persistent currents and their lack of electrical resistance, confer numerous benefits in terms of sensitivity, efficiency, and speed. Superconducting electronics have various significant applications, which include:

1.8.1.1. Superconducting Quantum Interference Devices (SQUIDs)

SQUIDs, an acronym for superconducting quantum interfering devices, are extensively employed within the realm of superconducting devices. Superconducting Quantum Interference Devices (SQUIDs) exhibit remarkable sensitivity as magnetic field detectors, enabling them to measure magnetic fields accurately. The operational mechanism of the aforementioned system is rooted in the concept of quantum interference, whereby alterations in magnetic flux give rise to observable modifications in the superconducting state. Superconducting Quantum Interference Devices (SQUIDs) have been utilized in diverse fields such as geophysics, medical diagnosis, materials evaluation, and non-destructive testing. SQUIDs are utilized in medical imaging applications, specifically in magnetoencephalography (MEG) and biomagnetic measurements. These techniques significantly contribute to understanding brain function and diverse physiological mechanisms (Fevrier & Laumond, 1986).

1.8.1.2. Superconducting Digital Circuits and Qubits

Superconducting digital circuits leverage the zero-resistance characteristics of superconductors to attain high-speed and energy-efficient computation. The circuits in question are founded upon Josephson junctions, which rely on the superconductive characteristics of slender insulating layers that are interposed between superconductive substances. Josephson junctions enable superconducting digital circuits to attain clock With higher frequencies in the gigahertz that vary and function at remarkably low power levels. The aforementioned circuits are utilized in various domains, including but not limited to high-speed processing of data, processing of digital signals, communication systems, and radar technology (Panek, Prando, & ALI MANSOORI, 1988).

Superconducting electronics encompass superconducting qubits, which are the fundamental units of quantum computers. Qubits are the quantum counterparts of classical bits and can exist in superposition states, thereby facilitating quantum parallelism and increasing computational prowess. Josephson junctions are commonly employed in implementing superconducting qubits, wherein the superconducting circuit's energy levels and coherence characteristics represent the quantum state. Despite being in the research and development phase, superconducting qubits exhibit significant potential for facilitating feasible quantum computers that can solve intricate problems efficiently.

1.8.2. Superconducting Magnets

Superconductors have found significant application in the form of superconducting magnets, which have brought about a revolutionary impact in diverse fields that depend on robust magnetic fields. The utilization of superconducting materials in magnet technology enables the production of elevated magnetic fields while minimizing energy consumption. Superconducting magnets have been utilized in various significant applications, such as:

1.8.2.1. Outline of Superconducting Magnet Applications

Superconducting magnets have a broad range of scientific, medical, and commercial applications. Charged particle guidance and focusing is a crucial application of these devices in particle physics research, particularly in particle accelerators like the Large Hadron Collider (LHC). Superconducting magnets are essential components in magnetic resonance imaging (MRI) systems, as they generate robust and consistent magnetic fields that are imperative for achieving high-quality medical imaging with exceptional resolution. Additional applications encompass magnetic separation methodologies, nuclear magnetic resonance (NMR) spectroscopic analysis, and fusion investigations (Figure 1.7).

Figure 1.7. *Image of compact muon solenoid detector on the LHC.*

Source: CERN, Creative Commons License.

Activity 1.1

Imagine you are organizing a science fair for high school students. Design an interactive demonstration that showcases the properties of superconductors and allows students to observe their unique behavior firsthand.

1.8.2.2. Magnetic Resonance Imaging (MRI) and NMR Systems

Superconducting magnets constitute the fundamental component of Magnetic Resonance Imaging (MRI) systems, facilitating the acquisition of intricate images of the human anatomy for diagnostic objectives. The precise imaging of the body is facilitated by the alignment of nuclear spins of atoms through the production of a robust magnetic field by the superconducting magnet. Utilizing superconducting magnets in magnetic resonance imaging (MRI) systems yields enhanced image resolution, expedited scanning durations, and heightened patient satisfaction relative to traditional resistive magnets. Nuclear magnetic resonance (NMR) systems utilized in scientific studies and chemical examination depend on superconducting magnets to produce the requisite magnetic fields for investigating the conduct of atomic nuclei (Leith et al., 2021).

1.8.3. Power and Energy Uses

Superconductors can transform power and energy systems by substantially augmenting efficacy, diminishing energy dissipation, and facilitating condensed and lightweight resolutions. Several important uses in the power and energy industries encompass:

1. **Superconducting Power Transfer and Distribution:** It is a field of study that uses materials that exhibit zero electrical resistance at low temperatures to transmit and distribute electrical power. Superconducting materials have the potential to enable the transmission of electrical power with negligible losses. Conventional power transmission lines are subject to considerable resistive losses, particularly when spanning extensive distances, leading to energy dissipation and reduced efficacy. In contrast, superconductivity power cables can transport substantial electrical current with negligible resistance, reducing energy losses. The aforementioned technology can transform power distribution networks, facilitating extended-range transmission while mitigating environmental repercussions and enhancing

overall efficacy. Superconducting current fault limiters represent a notable application within this field, offering expeditious and efficient safeguarding against faults in power systems (Hassenzahl et al., 2004).

2. **Superconducting Energy Storage Systems:** The energy storage is a crucial element in contemporary power systems, facilitating effective load management, peak demand reduction, and assimilation of sustainable energy resources. SMES (superconducting energy storage systems) use superconducting materials' capacity to store electrical energy in the form of ongoing currents. The aforementioned systems can expeditiously store and discharge energy, rendering them highly suitable for deployment in scenarios necessitating elevated power outputs, including stabilising electrical grids and providing supplementary power during periods of heightened demand. Small and medium-sized energy storage (SMES) systems exhibit a notable energy density and minimal energy dissipation, rendering them a potentially viable option for fulfilling forthcoming energy storage demands. Furthermore, superconducting magnetic energy storage (SMES) systems offer an alternative method for energy storage through the utilization of robust magnetic fields produced by superconducting magnets (Mitani, Tsuji, & Murakami, 1988).

1.8.4. Other Applications and Future Directions

Superconductors have been utilized in various fields, including electronics, magnets, and power systems. Current research and development endeavors are investigating diverse domains in which superconductivity can offer distinctive benefits. Several possible uses and future directions have been identified, such as:

1. **Superconducting Sensors and Detectors:** They are highly sensitive and accurate instruments with proven valuable assets in various fields, including astronomy, materials analysis, tracking the environment, and security screening. Superconducting transition-edge sensors (TES) can detect minute variations in temperature, radiation, and magnetic fields, facilitating accurate measurements and recognizing signals at the quantum level. The sensors in question have a wide range of practical uses, including but not limited to their deployment in astrophysics, particle

physics tests, non-destructive assessment methods, and national security systems (Oyama et al., 2008).

2. **Quantum Computing and Other Emerging Technologies:** The foundation of quantum computers is established by superconducting qubits, which are among the emerging technologies in the field of quantum computing. Quantum computing utilizes the basic tenets of quantum mechanics to execute computations that exhibit exponential acceleration compared to classical computers, specifically for particular applications. Superconducting qubits are considered one of the most promising candidates in the pursuit of developing viable quantum computers due to their extended coherence times and ability to be integrated with current fabrication methods. Furthermore, superconductors are currently being investigated in various nascent technologies, including superconducting spintronics, superconducting nanowires utilized in quantum communication, and superconducting metamaterials employed in sophisticated electromagnetic applications (Weinstock, 2000).

The domain of superconductivity is progressively evolving, with persistent scholarly investigations aimed at identifying fresh superconductive substances, enhancing crucial characteristics, and investigating innovative utilities. As comprehension of superconductivity progresses and manufacturing methods improve, the possibility of superconductors transforming diverse sectors and facilitating innovative technologies grows more auspicious. Sustained allocation of resources towards research and development, coupled with synergistic partnerships among academia, industry, and government, will be instrumental in fully harnessing the capabilities of superconductors in pragmatic contexts (Schneider, 1991).

1.9. CHALLENGES AND LIMITATIONS

1.9.1. Current Challenges in Superconductivity

1.9.1.1. Boundaries of Existing Superconducting Materials

The advancement of superconducting materials has been noteworthy; however, certain constraints continue to impede their extensive implementation and pragmatic use. A significant obstacle pertains to the temperature range within which superconductivity manifests. Conventional superconductors necessitate cryogenic cooling to achieve the critical temperature (Tc), which is in close proximity to absolute zero (-273.15°C). Conversely, high-temperature superconductors (HTS) possess a relatively higher Tc, yet they still necessitate cryogenic cooling. The aforementioned constraint presents pragmatic challenges and curtails the attainability of superconductivity in commonplace implementations (Yazdani-Asrami, Seyyedbarzegar, Sadeghi, de Sousa, & Kottonau, 2022).

A further constraint pertains to superconductive substances' critical current density (Jc). The critical current density (Jc) refers to the upper limit of electrical current density that a given material can sustain while retaining its superconducting properties. Several superconductors demonstrate a distinct decline in Jc when subjected to intense magnetic fields or when strong currents are present. The critical current loss phenomenon imposes limitations on the practicality of superconductors in scenarios that necessitate high current-carrying capabilities or exposure to potent magnetic fields (Zhang et al., 2019).

Moreover, the production and manipulation of superconducting substances present certain difficulties. Certain superconductors pose challenges in terms of synthesis and may necessitate intricate manufacturing procedures, resulting in significant time and cost expenditures. Furthermore, the technical challenge of producing long wires or massive superconducting establishes with high consistency and quality continues to impede superconducting devices' scalability and commercial feasibility (Sun & Cava, 2019).

1.9.1.2. Cost and Scalability Considerations

Cost considerations significantly impact the widespread use of superconducting technologies. Fabricating superconducting materials frequently entails utilizing scarce or costly elements, intricate production techniques, and specialized cryogenic cooling apparatus. The aforementioned factors significantly contribute to the elevated production expenses associated with superconductors. Consequently, superconducting technologies are presently confined to specialized applications that warrant substantial costs, such as in the domains of medical imaging or accelerators for particles (Schwartz et al., 2008).

The issue of scalability presents a significant obstacle in the realm of superconductivity. Although laboratory-scale demonstrations and prototypes have exhibited encouraging outcomes, upscaling these technologies to industrial or commercial levels is still challenging. The amalgamation of superconducting materials with pre-existing infrastructure and traditional methods presents a set of engineering hurdles. The development of cooling systems that are both efficient and cost-effective for large-scale applications, while additionally guaranteeing compatibility with other components can be a complex undertaking that necessitates a great deal of research efforts (Baecker, 2011).

1.9.2. Future Directions and Predictions

1.9.2.1. Ongoing Research and Advancements

Notwithstanding the extant challenges, persistent research and progress in the field of superconductivity persist in confronting the constraints and charting a course for forthcoming innovations. Researchers in the fields of science and engineering are currently investigating diverse methodologies to augment the efficiency and broaden the applicability of superconducting materials (Kroemer, 2003).

An active area of research pertains to exploring and advancing novel categories of superconductors that exhibit elevated critical temperatures. This could facilitate the occurrence of superconductivity under less stringent cooling conditions, potentially obviating the requirement for costly cryogenic systems. Furthermore, ongoing endeavors are being made to enhance the critical current density of superconductors, thereby augmenting their resistance to external magnetic fields and enabling them to transport greater magnitudes of electrical current (Flores-Livas et al., 2020).

An additional avenue of research pertains to the investigation of alternative methods of fabrication and innovative materials. Researchers endeavor to devise more efficient and economical techniques for fabricating superconducting materials by exploring various synthesis approaches, such as chemical depositing, thin film methods, and nanoscale engineering. In addition, the progress made in the field of material science and nanotechnology has the potential to unveil novel categories of superconductors that exhibit distinctive characteristics and enhanced efficacy. Scholars also focus on comprehending the fundamental mechanisms that underlie superconductivity. The investigation examines the functions of electron-phonon conversations, spin fluctuations, and atypical pairing mechanisms within various categories of superconductors. Through further exploration and analysis of these phenomena, researchers aim to uncover novel avenues for advancing superconductor technology, characterized by improved properties and elevated operational temperatures (Ford & Saunders, 2004).

1.9.2.2. Potential Breakthroughs and Applications

The potential advancements and diverse applications of superconductivity present a promising outlook for the future in various industries. Several potential domains of influence may be considered, including:

1. **Power Generation and Transmission:** The implementation of superconducting materials has the potential to bring about a significant transformation in power generation, distribution, and transmission systems. Implementing lossless power transmission would facilitate the effective conveyance of electricity across extensive distances, thereby mitigating energy losses and augmenting the grid's stability. The implementation of superconducting current fault limiters has the potential to mitigate grid failures that occur during faults, thereby enhancing the overall dependability of the power infrastructure (MacManus-Driscoll & Wimbush, 2011).
2. **Energy Storage:** Superconducting energy storage systems can store substantial quantities of energy while incurring negligible losses. Implementing such systems can potentially improve the integration of renewable energy sources, thereby ensuring a consistent and dependable electricity supply even during intermittent generation periods. Superconducting magnetic energy storage (SMES) systems, which utilize superconducting coils, can expeditiously release stored energy as required, rendering them appropriate for high-power applications.
3. **Transportation:** The implementation of superconductivity has the potential to revolutionize transportation systems by developing propulsion systems that are energy-efficient, lightweight, and capable of generating high power. Utilizing superconducting motors and magnetic levitation (maglev) trains presents a promising prospect for achieving high-speed and energy-efficient transportation, characterized by minimal friction and reduced environmental footprint.
4. **Quantum Computing:** The potential of superconducting qubits, fundamental units of quantum computers, to address intricate computational issues is highly encouraging. Further investigation and progress in the field of superconducting quantum computing may result in noteworthy advancements in cryptography, optimizing, and simulation (Shokair et al., 2014).

5. **Healthcare and Scientific Instruments:** Superconducting magnets have been extensively employed in medical imaging equipment, including magnetic resonance imaging (MRI) and nuclear magnetic resonance (NMR) spectrometers, within the healthcare and scientific instruments industry. Advancements in superconducting substances and magnet technology can augment the resolution and capabilities of diagnostic and research instruments.
6. **Other Applications:** Superconducting materials are utilized in diverse fields, such as particle accelerators, sensors, detectors, and communication systems. Continued research and technological progress have the potential to result in innovative applications and enhanced performance within these domains (Alivisatos et al., 1998).

1.10. SUMMARY

Chapter one provides an overview of superconductors, starting with an introduction to the topic and a historical background of superconductivity. It highlights key milestones in superconductivity research and the contributions of prominent scientists. The chapter explores the features and properties of superconductors, including zero electrical resistance, persistent currents, perfect diamagnetism, and the Meissner effect. It also discusses the critical magnetic field and critical current density. The chapter categorizes superconductors into Type I and Type II, and introduces the BCS theory of superconductivity. It delves into high-temperature superconductivity, covering challenges, findings, proposed concepts, and notable high-temperature superconductors. The chapter further explores LTS, their common materials, properties, and fabrication practices. It concludes by discussing the various uses of superconductors, such as in electronics, magnets, power and energy applications, and provides an overview of current challenges, future directions, and predictions in the field.

REVIEW QUESTIONS

1. What is the historical significance of the discovery of superconductivity?
2. Name some prominent scientists and their contributions to the field of superconductivity.
3. What are the key features and properties of superconductors?
4. Explain the concepts of zero electrical resistance and persistent currents.
5. What is the Meissner effect and how does it relate to superconductivity?
6. Define critical magnetic field and critical current density in superconductors.
7. Differentiate between Type I and Type II superconductors.
8. What is the BCS theory of superconductivity and how does it explain the phenomenon?
9. What are the challenges and findings in high-temperature superconductivity?
10. Discuss some of the proposed concepts and mechanisms for high-temperature superconductivity.
11. Provide an outline of notable high-temperature superconductors.
12. Describe the common materials and fabrication practices for low-temperature superconductors.
13. How are superconductors used in electronics, magnets, power, and energy applications?
14. Discuss some other potential applications and future directions for superconductors.
15. What are the current challenges in superconductivity research?
16. What are some predictions and future directions for the field of superconductivity?

REFERENCES

1. Alivisatos, P., Barbara, P. F., Castleman, A. W., Chang, J., Dixon, D. A., Klein, M. L., Rossky, P. J. (1998). From molecules to materials: Current trends and future directions. *Advanced Materials, 10*(16), 1297–1336.
2. Allen, P. B., & Mitrović, B. (1983). Theory of superconducting Tc. *Solid State Physics, 37*, 1–92.
3. Anlage, S. M. (2010). The physics and applications of superconducting metamaterials. *Journal of Optics, 13*(2), 024001.
4. Babaev, E., & Speight, M. (2005). Semi-Meissner state and neither type-I nor type-II superconductivity in multicomponent superconductors. *Physical Review B, 72*(18), 180502.
5. Baecker, M. (2011). Energy and superconductors–applications of high-temperature-superconductors.
6. Balian, R., Flocard, H., & Vénéroni, M. (1999). Variational extensions of BCS theory. *Physics Reports, 317*(5–6), 251–358.
7. Bardeen, J., Cooper, L. N., & Schrieffer, J. R. (1957). Theory of superconductivity. *Physical Review, 108*(5), 1175.
8. Bednorz, J. G., & Müller, K. A. (1986). Possible high T c superconductivity in the Ba–La–Cu–O system. *Zeitschrift für Physik B Condensed Matter, 64*(2), 189–193.
9. Bednorz, J. G., & Müller, K. A. (1988). Perovskite-type oxides—the new approach to high-Tc superconductivity. *Reviews of Modern Physics, 60*(3), 585.
10. Bertrand, D. (2005). A relativistic BCS theory of superconductivity. *An Experimentally Motivated Study of Electric Fields in Superconductors, Université catholique de Louvain.*
11. Bugoslavsky, Y., Cohen, L., Perkins, G., Polichetti, M., Tate, T., Gwilliam, R., & Caplin, A. (2001). Enhancement of the high-magnetic-field critical current density of superconducting MgB_2 by proton irradiation. *Nature, 411*(6837), 561–563.
12. Burns, G. (1992). *High-Temperature Superconductivity, 1(1), 3-20.*
13. Chancellor, N., Zohren, S., & Warburton, P. A. (2017). Circuit design for multi-body interactions in superconducting quantum annealing systems with applications to a scalable architecture. *npj Quantum Information, 3*(1), 21.
14. Chubukov, A. (2012). Pairing mechanism in Fe-based superconductors. *Annu. Rev. Condens. Matter Phys., 3*(1), 57–92.

15. Clarke, J. (1968). Experimental comparison of the Josephson voltage-frequency relation in different superconductors. *Physical Review Letters, 21*(23), 1566.
16. Dagotto, E. (1994). Correlated electrons in high-temperature superconductors. *Reviews of Modern Physics, 66*(3), 763.
17. Dew-Hughes, D. (1974). Flux pinning mechanisms in type II superconductors. *Philosophical Magazine, 30*(2), 293–305.
18. Emery, V., & Reiter, G. (1988). Mechanism for high-temperature superconductivity. *Physical Review B, 38*(7), 4547.
19. Essén, H., & Fiolhais, M. C. (2012). Meissner effect, diamagnetism, and classical physics—a review. *American Journal of Physics, 80*(2), 164–169.
20. Fevrier, A., & Laumond, Y. (1986). *Prospective Uses of Superconductors for 50/60 Hertz Applications.* Paper presented at the Proceedings of the eleventh international cryogenic engineering conference.
21. Flores-Livas, J. A., Boeri, L., Sanna, A., Profeta, G., Arita, R., & Eremets, M. (2020). A perspective on conventional high-temperature superconductors at high pressure: Methods and materials. *Physics Reports, 856*, 1–78.
22. Ford, P. J., & Saunders, G. A. (2004). *The Rise of the Superconductors*: CRC Press.
23. Ginzburg, V. L., Kirzhnits, D. A., Agyei, A., & Birman, J. L. (1982). *High-Temperature Superconductivity*: Springer.
24. Goodman, B. (1966). Type II superconductors. *Reports on Progress in Physics, 29*(2), 445.
25. Gropp, W. D., Kaper, H. G., Leaf, G. K., Levine, D. M., Palumbo, M., & Vinokur, V. M. (1996). Numerical simulation of vortex dynamics in type-II superconductors. *Journal of Computational Physics, 123*(2), 254–266.
26. Halperin, B. I., Refael, G., & Demler, E. (2010). Resistance in superconductors. *International Journal of Modern Physics B, 24*(20n21), 4039–4080.
27. Hassenzahl, W. V., Hazelton, D. W., Johnson, B. K., Komarek, P., Noe, M., & Reis, C. T. (2004). Electric power applications of superconductivity. *Proceedings of the IEEE, 92*(10), 1655–1674.
28. Heine, K., Tenbrink, J., & Thöner, M. (1989). High-field critical current densities in $Bi_2Sr_2Ca_1Cu_2O_8^+$ x/Ag wires. *Applied Physics Letters, 55*(23), 2441–2443.
29. Hirsch, J. (2009). BCS theory of superconductivity: it is time to question its validity. *Physica Scripta, 80*(3), 035702.

30. Hirsch, J. (2016). On the dynamics of the Meissner effect. *Physica Scripta, 91*(3), 035801.
31. Hott, R. (2003). Materials aspects of high-temperature superconductors for applications. *arXiv preprint cond-mat/0306442.*
32. Jackson, T., & Palmer, S. (1994). Oxide superconductor and magnetic metal thin film deposition by pulsed laser ablation: a review. *Journal of Physics D: Applied Physics, 27*(8), 1581.
33. Johnston, D. C. (2013). Elaboration of the α-model derived from the BCS theory of superconductivity. *Superconductor Science and Technology, 26*(11), 115011.
34. Kordyuk, A. (2015). Pseudogap from ARPES experiment: three gaps in cuprates and topological superconductivity. *Low Temperature Physics, 41*(5), 319–341.
35. Krabbes, G., Fuchs, G., Canders, W.-R., May, H., & Palka, R. (2006). *High Temperature Superconductor Bulk Materials*: Wiley-Vch Weinheim.
36. Kroemer, H. (2003). Speculations about future directions. *Journal of Crystal Growth, 251*(1–4), 17–22.
37. Lancaster, M. J. (1997). Passive microwave device applications of high-temperature superconductors, 1(2), 5-23.
38. Larbalestier, D., & Lee, P. (1995). *New developments in niobium titanium superconductors.* Paper presented at the Proceedings Particle Accelerator Conference.
39. Lee, P. J., & Larbalestier, D. C. (2003). Niobium-titanium superconducting wires: Nanostructures by extrusion and wiredrawing. *Wire Journal International, 36*(2), 61–66.
40. Leith, S., Vogel, M., Fan, J., Seiler, E., Ries, R., & Jiang, X. (2021). Superconducting NbN thin films for use in superconducting radio frequency cavities. *Superconductor Science and Technology, 34*(2), 025006.
41. Little, W. A. (1967). Decay of persistent currents in small superconductors. *Physical Review, 156*(2), 396.
42. London, F., & London, H. (1935). The electromagnetic equations of the supraconductor. *Proceedings of the Royal Society of London. Series A-Mathematical and Physical Sciences, 149*(866), 71–88.
43. Lozovik, Y. E., & Yudson, V. (1976). A new mechanism for superconductivity: pairing between spatially separated electrons and holes. *Zh. Eksp. Teor. Fiz, 71*, 738.
44. Ma, Y. (2012). Progress in wire fabrication of iron-based superconductors. *Superconductor Science and Technology, 25*(11), 113001.

45. MacManus-Driscoll, J. L., & Wimbush, S. C. (2011). Future directions for cuprate conductors. *IEEE Transactions on Applied Superconductivity, 21*(3), 2495–2500.

46. Mele, P., Prassides, K., Tarantini, C., Palau, A., Badica, P., Jha, A. K., & Endo, T. (2019). *Superconductivity: From Materials Science to Practical Applications*. Springer Nature.

47. Mitani, Y., Tsuji, K., & Murakami, Y. (1988). Application of superconducting magnet energy storage to improve power system dynamic performance. *IEEE Transactions on Power Systems, 3*(4), 1418–1425.

48. Mitschang, G. W. (1995). Space applications and implications of high temperature superconductivity. *IEEE Transactions on Applied Superconductivity, 5*(2), 69–73.

49. Müller, K. A., & Bednorz, J. G. (1987). The discovery of a class of high-temperature superconductors. *Science, 237*(4819), 1133–1139.

50. Onnes, M. H., Leiden, C., & Onnes, H. K. (2016). What are superconductors. *Renewable and Sustainable Energy Reviews, 55*, 59–72.

51. Orenstein, J., & Millis, A. (2000). Advances in the physics of high-temperature superconductivity. *Science, 288*(5465), 468–474.

52. Oyama, H., Shinzato, T., Hayashi, K., Kitajima, K., Ariyoshi, T., & Sawai, T. (2008). Application of superconductors for automobiles. *SEI Tech. Rev, 67*, 22–26.

53. Panek, D., Prando, G., & ALI MANSOORI, G. (1988). Superconductors: Past, present, and future. *Energy Sources, 10*(3), 159–172.

54. Paul, W., Chen, M., Lakner, M., Rhyner, J., Braun, D., & Lanz, W. (2001). Fault current limiter based on high temperature superconductors – different concepts, test results, simulations, applications. *Physica C: Superconductivity, 354*(1–4), 27–33.

55. Petrovic, C., Movshovich, R., Jaime, M., Pagliuso, P., Hundley, M., Sarrao, J., Thompson, J. (2001). A new heavy-fermion superconductor $CeIrIn_5$: A relative of the cuprates? *Europhysics Letters, 53*(3), 354.

56. Pippard, A. (1947). The surface impedance of superconductors and normal metals at high frequencies i. resistance of superconducting tin and mercury at 1200 mcyc./sec. *Proceedings of the Royal Society of London. Series A. Mathematical and Physical Sciences, 191*(1026), 370–384.

57. Prozorov, R., Fidler, A. F., Hoberg, J. R., & Canfield, P. C. (2008). Suprafroth in type-I superconductors. *Nature Physics, 4*(4), 327–332.

58. Qiu, X. G. (2011). *High-Temperature Superconductors*. Elsevier, 2(1), 4-8.

59. Ren, Z. A., & Zhao, Z. X. (2009). Research and Prospects of Iron-Based Superconductors. *Advanced Materials, 21*(45), 4584–4592.

60. Rosenstein, B., & Li, D. (2010). Ginzburg-Landau theory of type II superconductors in magnetic field. *Reviews of Modern Physics, 82*(1), 109.

61. ŞAKİROĞLU, S. (2002). *Investigation of Various Substitution Effects on BiPbSrCaCuO Type High Temperature Ceramic Superconductors.* DEÜ Fen Bilimleri Enstitüsü.

62. Saxena, A. K. (2012). *High-Temperature Superconductors* (Vol. 125): Springer Science & Business Media.

63. Schmalian, J. (2010). Failed theories of superconductivity. *Modern Physics Letters B, 24*(27), 2679–2691.

64. Schneider, T. R. (1991). Energy applications of superconductivity. *Annual Review of Energy and the Environment, 16*(1), 533–555.

65. Schrieffer, J. R. (2018). *Theory of Superconductivity*. CRC Press, 1(3), 9-33.

66. Schwartz, J., Effio, T., Liu, X., Le, Q. V., Mbaruku, A. L., Schneider-Muntau, H. J., Wang, X. (2008). High field superconducting solenoids via high temperature superconductors. *IEEE Transactions on Applied Superconductivity, 18*(2), 70–81.

67. Shein, I., & Ivanovskii, A. (2009). Electronic and structural properties of low-temperature superconductors and ternary pnictides A Ni_2 Pn_2 (A= Sr, Ba and P n= P, As). *Physical Review B, 79*(5), 054510.

68. Shokair, T., Root, J., Van Bibber, K., Brubaker, B., Gurevich, Y., Cahn, S., Mitchell, B. (2014). Future directions in the microwave cavity search for dark matter axions. *International Journal of Modern Physics A, 29*(19), 1443004.

69. Smith, T., Chu, C., & Maple, M. (1969). Superconducting manometers for high pressure measurement at low temperature. *Cryogenics, 9*(1), 53–56.

70. Sonier, J. E., Brewer, J. H., & Kiefl, R. F. (2000). μSR studies of the vortex state in type-II superconductors. *Reviews of Modern Physics, 72*(3), 769.

71. Strongin, M., Paskin, A., Schweitzer, D. G., Kammerer, O., & Craig, P. (1964). Surface superconductivity in type I and type II superconductors. *Physical Review Letters, 12*(16), 442.

72. Suhl, H., Matthias, B., & Walker, L. (1959). Bardeen-Cooper-Schrieffer theory of superconductivity in the case of overlapping bands. *Physical Review Letters, 3*(12), 552.

73. Sun, L., & Cava, R. J. (2019). High-entropy alloy superconductors: Status,

opportunities, and challenges. *Physical Review Materials, 3*(9), 090301.

74. Thieme, C., Pourrahimi, S., Schwartz, B., & Foner, S. (1984). Improved high field performance of Nb-Al powder metallurgy processed superconducting wires. *Applied Physics Letters, 44*(2), 260–262.

75. van der Laan, D. C., Ekin, J., Douglas, J., Clickner, C., Stauffer, T. C., & Goodrich, L. (2010). Effect of strain, magnetic field and field angle on the critical current density of $Y Ba_2Cu_3O_{7-\delta}$ coated conductors. *Superconductor Science and Technology, 23*(7), 072001.

76. Weinstock, H. (2000). *Applications of Superconductivity* (Vol. 365): Springer Science & Business Media.

77. Wesche, R. (2015). *Physical Properties of High-Temperature Superconductors*: John Wiley & Sons, 2(4), 5-9.

78. Yamaguchi, K., Hata, Y., Mochiku, T., & Yasuoka, H. (2012). Optimization of annealing processes on superconductivity of $FeSr_2YCu_2O_{6+\delta}$. *Physics Procedia, 27*, 24–27.

79. Yazdani-Asrami, M., Seyyedbarzegar, S., Sadeghi, A., de Sousa, W. T., & Kottonau, D. (2022). High temperature superconducting cables and their performance against short circuit faults: current development, challenges, solutions, and future trends. *Superconductor Science and Technology, 35*(8), 083002.

80. Zhang, H., Xiao, L., Luo, B., Guo, J., Zhang, L., & Xie, J. (2019). The potential and challenges of time-resolved single-photon detection based on current-carrying superconducting nanowires. *Journal of Physics D: Applied Physics, 53*(1), 013001.

81. Zhou, X., Lee, W.-S., Imada, M., Trivedi, N., Phillips, P., Kee, H.-Y., Eremets, M. (2021). High-temperature superconductivity. *Nature Reviews Physics, 3*(7), 462–465.

CHAPTER

2

General Theory of High-T_C Superconductors

Contents

2.1. INTRODUCTION

The general theory ofhigh-Tc superconductors is a comprehensive framework that seeks to elucidate the fundamental concepts governing high-temperature superconducting materials' behavior. The high-Tc superconductors exhibit superconductivity at temperatures that surpass those of conventional superconductors. The field of superconductivity initially focused on low-temperature phenomena, but the emergence of high-Tc superconductors in the late 1980s opened up new prospects for practical utilization (Wang & Lee, 2003).

The integration of various theoretical frameworks and empirical observations in the General Theory of high-Tc superconductors justifies the unique properties exhibited by these materials. The lack of a widely recognized theory presents a considerable obstacle to understanding high-temperature superconductivity. Several theoretical frameworks have been proposed, such as the BCS-BEC crossover, d-wave pairing theory, and resonating valence bond theory. The theories above aim to explain the unconventional pairing mechanisms and importance of interactions of electron-electron in high-temperature superconductors. The primary aim of the General Theory of high-Tc superconductors is to synthesize various perspectives and achieve a comprehensive understanding of the complex phenomena that occur in these materials. The outcome of this endeavor is the progression of superconducting technologies (Fradkin et al., 2015).

2.2. FUNDAMENTAL PROPERTIES OF SUPERCONDUCTORS

A few properties of superconductors are discussed in subsections.

2.2.1. Superconducting Phenomenon

The gradual degradation of the superconductor's characteristics as temperature decreases ultimately leads to a sudden cessation. Critical temperature, symbolized as Tc, is temperature where this phenomenon occurs (Herbut, 2005).

Tip

Understanding the quantum mechanical principles underlying superconductivity is essential for comprehending the behavior of high-TC superconductors.

Process of water transitioning to ice at its freezing point involves identifying discrete phases, specifically the liquid phase of water and the solid phase of ice. Superconductivity can be categorized as a phase transition phenomenon, which involves the observation of a transition between two distinct phases. When the temperature surpasses critical temperature (Tc), the superconductor transitions to a normal conduction characterized by electrical resistance. Electron system lacks order within the framework, as denoted by the mathematical expression $\psi = 0$. By contrast, superconductor undergoes a transition into a superconducting state and exhibits observable electron system order when cooled below its critical temperature (Tc) (Kruchinin, 2014).

Then, $\psi \neq 0$. The phenomenon of Cooper pairing leads to the creation of Cooper pairs, which are plentiful in the superconductive material upon decreasing temperature. The pairs above are characterized by wave functions that exhibit equivalent amplitude and phase. As a result, they intersect with one another, leading to the emergence of a wave function that can be detected on a macroscopic level. The quantity denoted as ψ is the determining factor in calculating the macroscopic wave function. In cases where ψ is not equal to zero, the electrical resistance experiences a significant decrease, and a range of superconducting phenomena, including but not limited to the persistent current, Meissner effect, and Josephson effect, become apparent (Kivelson & Fradkin, 2005) (Figure 2.1).

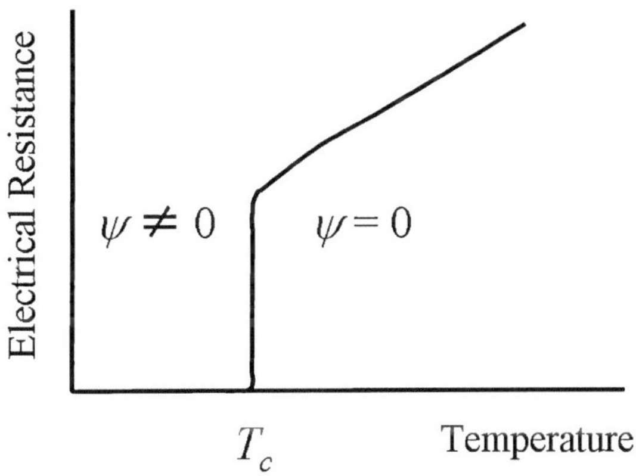

Figure 2.1. Typical temperature dependence of an electrical resistance of a superconductor. T_c is the critical temperature. When $\psi \neq 0$ ($T < T_c$) the material is in the superconducting state, whereas; $\psi = 0$ ($T > T_c$) corresponds to the normal conducting state.

Source: Kaname, Creative Commons License.

The process of thermal dissipation arising from the electrical resistance of a metallic substance, such as copper or aluminum, is evident upon the flow of an electric current. The energy dissipation in an electric power cable is a noteworthy phenomenon that cannot be disregarded. Applying a significant electrical current to a copper coil with the intention of producing a strong magnetic field can lead to the melting of the copper wire due to the consequential thermal energy. In contrast, when a superconductor is exposed to a current above its critical temperature (Tc), it does not dissipate energy due to its absence of electrical resistance. Utilizing a superconducting coil facilitates the production of a potent magnetic field, thereby facilitating power transmission through a cable with zero energy dissipation (Emery & Kivelson, 1993).

Figure 2.2 portrays a graphical representation showcasing critical temperature Tc advancement to superconductors. Kamerlingh-Onnes discovered mercury superconductivity in 1911. Successive discoveries of superconductive materials followed, gradually elevating critical temperature (Tc). The cuprate superconductors emergence can be traced back to the year 1986, and after this event, there has been a notable surge in Tc. At an elevated pressure of 30 gigapascals, the critical temperature (Tc) was observed to attain a value of 164°K, as reported in reference. This suggests that maximum threshold of Tc had already reached 50% of the typical ambient temperature (around 300 K) (Franke et al., 2011).

The substances mentioned above are frequently denoted as high-temperature superconductors in academic literature. The unearthing of MgB_2, a metallic superconductor which exhibits critical temperature (T_c) of 39 K, in the year 2001 has attracted considerable interest in contemporary times. This subject matter has been extensively deliberated within scholarly communities. New class of materials was discovered in 2008, demonstrating a T_c value of around 56 K. The materials under consideration were categorized as iron pnictides, which consist of iron and arsenide, as opposed to the previously identified high-temperature superconductors composed of copper and oxygen. The finding above has incited a revived fascination in superconductors with high critical temperatures (high-T_c) (Bergeret et al., 2005).

The pragmatic utilization of traditional superconducting wires composed of Nb-Ti and Nb_3Sn has been substantiated through the advancement of diverse technologies, including, magnetic resonance imaging (MRI), magnetic levitation trains, particle accelerators, nuclear magnetic resonance

(NMR), and other related fields. This provides evidence of the efficacy of these superconductors. Metallic wires are of utmost importance in manufacturing the high-capacity superconducting coil utilized in a nuclear fusion device. Nevertheless, exposing them to a cooling process at a temperature of 4.2°K is crucial, which requires the utilization of expensive liquid helium (Shurakoy et al., 2015).

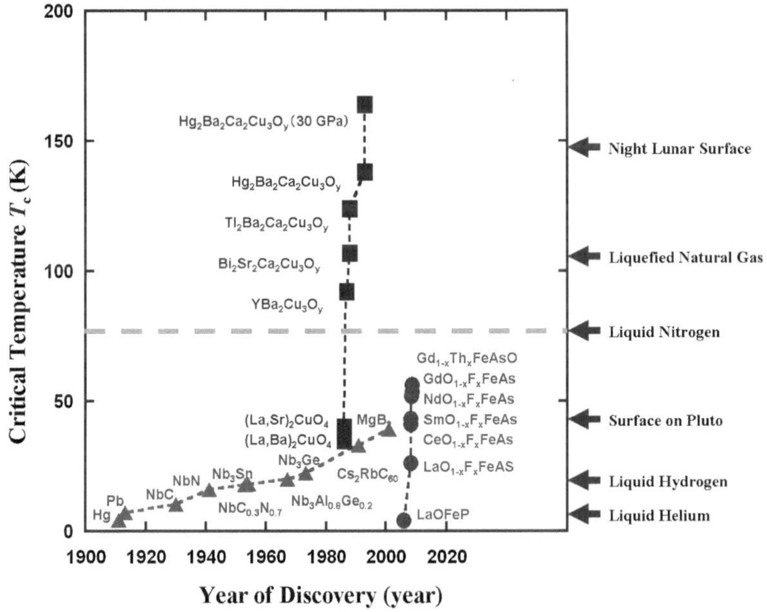

Figure 2.2. The evolution of the critical temperature T_c of superconductors. The series of high-T_c cuprate superconductors were first discovered in 1986. Iron arsenide superconductors, which are other candidates for high-T_c superconductors, were discovered in 2008.

Source: Matsumoto, Creative Commons License.

High-temperature superconductors have achieved a critical temperature exceeding 77 K, Achieving superconductivity by reducing electrical resistance in liquid nitrogen. A copious nitrogen amount in atmosphere facilitates a noteworthy decrease in cooling expenditures. This can be attributed to the cost of liquid nitrogen being significantly lower than that of liquid helium. The implementation of high-Tc superconductors in electric power and other industrial technologies is anticipated to substantially improve the efficacy of electrical energy consumption. Identifying a superconductor that operates at room temperature, thereby displaying superconductivity under ambient conditions, would eliminate liquid nitrogen's need to achieve complete electrical conductivity. This significant advancement has the potential to enable numerous applications in the realm of superconductivity. The pursuit of materials that demonstrate high Tc values continues with considerable intensity (Buzea & Robbie, 2004).

2.2.2. Superconductivity and Quantum Mechanics

A prerequisite for comprehending the wave function ψ in the superconducting phase is familiarity with quantum mechanics. This article provides a concise explanation of the correlation between superconductivity and quantum mechanics (Likharev, 1983).

2.2.2.1. Wave-Particle Duality

A subatomic entity, exemplified by electron, demonstrates dualistic attributes of particle and wave nature. For instance, if mass particle is in motion with a velocity of v, it exhibits an energy of $E = (1/2)\ mv^2$ and momentum of . de Broglie equation establishes a correlation between momentum of particle and its wavelength λ, thereby indicating that a particle demonstrates wave-like characteristics (Afshar et al., 2007)

$$\lambda = h/p \tag{1}$$

Planck's constant is represented by the symbol "h." The concept of wave-particle duality is responsible for the emergence of the uncertainty principle, which establishes a correlation between the degree of uncertainty in the position (x) and time (t) of a given particle

$$\Delta x\ \Delta p \approx h,\ \Delta E\ \Delta t \approx h \tag{2}$$

The current investigation pertains to a particle restricted within a confined space of finite dimensions, specifically a length denoted by L. The wave-like behavior of the particle confined within the box is demonstrated by a standing wave with wavelengths of $\lambda = 2L$, L, and (2/3)L. The depiction of a wave's stationary state presented.

$$\psi(x) = A\sin(2\pi x\ \lambda)\exp(i\theta) \tag{3}$$

The variables θ and A represent phase of wave and the amplitude, respectively. An equation mentioned above can be differentiated twice using value of $\lambda = 2\ L/n$ (where n is a positive integer) to obtain the following equation (Dimitrova & Weis, 2008).

$$d^2\psi(x)/dx^2 = -(2\pi\lambda)^2\psi(x) \tag{4}$$

The Schrödinger equation, as presented in this work, is obtained by means of the de Broglie equation and the equation $E = (1/2)\,mv^2$. This leads to the reformulation of Eq. (4). It can be determined by defining h = h/ 2 π.

$$-({}^2/2m)d^2\psi/dx^2 = E\psi \qquad (5)$$

The mathematical construct is known as the wave function. The symbol ψ is widely recognized as a depiction of the wave-like characteristics exhibited by particles. The magnitude squared of the wave function, denoted as $|\psi|^2$, can be construed as wave amplitude from wave-centric viewpoint and as probability density of particle's existence from particle-centric viewpoint (Angelo & Ribeiro, 2015). Following this, particle energy takes on a discrete character and can be mathematically represented in following manner (Figure 2.3).

$$E = (h^2/2m)(2L/n)^2\,(n = 1, 2,\ldots) \qquad (6)$$

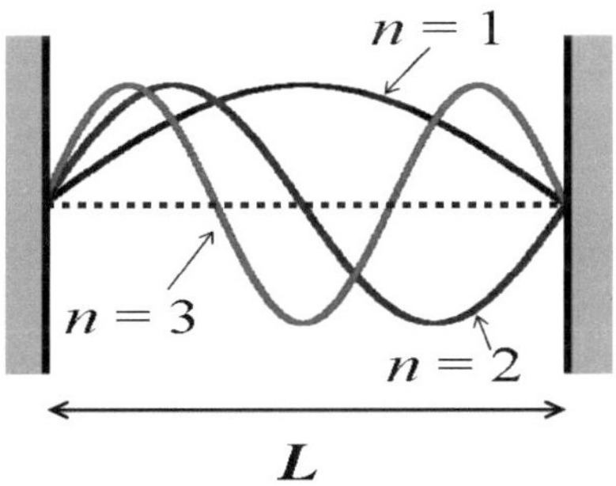

Figure 2.3. The microscopic particle-like electron which is confined in a small box shows wave-like properties. The Schrödinger equation predicts discrete energy levels.

Source: Matsumoto, Creative Commons License.

2.2.2.2. Fermion and Boson

The taxonomy of subatomic particles into fermions and bosons represents a noteworthy discovery in quantum mechanics. The particles mentioned above exhibit an intrinsic property referred to as spin. Fermions are subatomic particles that exhibit a half-integral spin, including electrons, protons, and neutrons. On the other hand, bosons, such as photons and phonons, possess integral spin, distinguishing them from fermions. The behavior of a composite particle composed of fermions depends on the total number of particles and the aggregate spin count. Specifically, fermionic behavior is observed when the number of particles is odd, while bosonic behavior is exhibited when the particles number is even., The atomic nucleus of helium-4 (^4He) is composed of two protons, two neutrons, and 2 electrons, manifesting bosonic characteristics. In contrast, the isotope helium-3 (^3He) is classified as fermion based on its constituent components of one neutron, 2 protons, and 2 electrons (Arndt et al., 1999).

The phenomenon of repulsion between fermions restricted within a confined spatial volume is attributed to the Pauli exclusion principle. It has been observed that a solitary energy level can accommodate only two fermions, which their positive and negative spins can distinguish. On the other hand, a plethora of bosonic particles can inhabit a common energy state (Kolesov et al., 2009) (Figures 2.4–2.6).

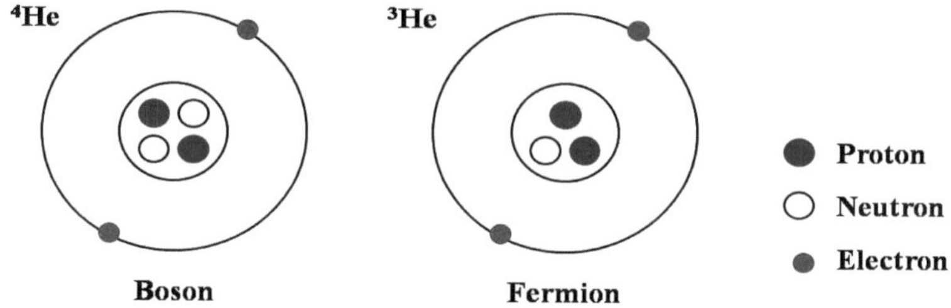

Figure 2.4. Particles composed of fermions exhibit either bosonic or fermionic behavior, which is determined by the aggregate number of spins. According to quantum mechanics, a helium-4 nucleus with six fermions is classified as a boson, while a helium-3 nucleus is classified as a fermion.

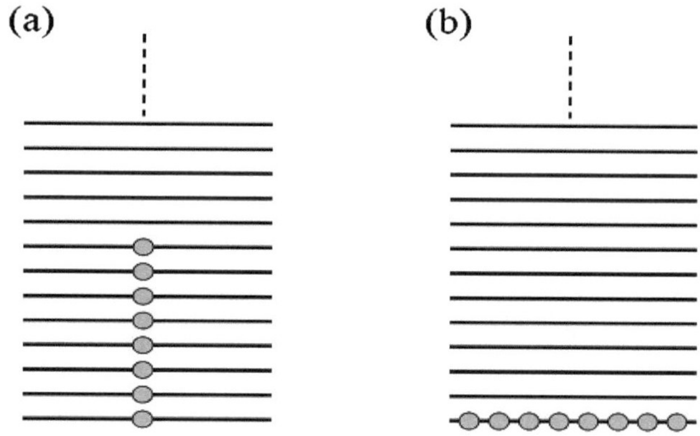

Figure 2.5. The topic of interest pertains to the distribution of subatomic particles within distinct energy levels when the temperature is zero. The particles depicted in (a) are fermions, while those in (b) are bosons. The arrangement of fermions occurs sequentially, beginning from lowest energy level and progressing toward the highest. Conversely, bosons are collectively placed within the lowest energy level.

Source: Kaname, Creative Commons License.

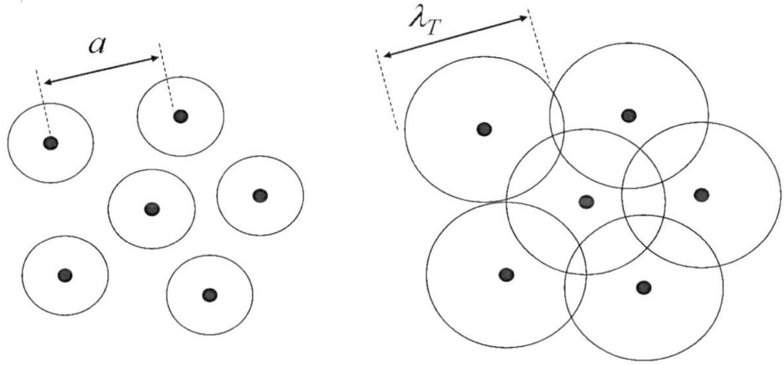

Figure 2.6. When the temperature is decreased, there is an overlapping of gas particles. When thermal de Broglie wavelength of gas particles, denoted by λ T, is approximately equal to the average interval between them, denoted by a, the particles begin to overlap. (a) depicts high temperature; while (b) represents low temperature.

Source: Kaname, Creative Commons License.

At point of absolute zero temperature, fermions display state of energy distribution where particles are ordered sequentially, beginning from lowest energy level and advancing one by one, as illustrated in Figure 2.5(a). In contrast, it can be observed that bosons display a unique state of energy distribution, whereby all particles are densely packed into lowest energy level, as depicted in Figure 2.5(b).

Based on an assumed average inter-particle distance denoted as "a" and a thermal de Broglie wavelength for gas particles, it's possible to draw the following conclusions (Hendry, 1980).

$\lambda_1 \approx h/(2\pi mk\ T_B)^{12}$ (7)

The equation described above pertains to the computation of mean kinetic energy of gas molecules. In this equation, the symbol kB represents Boltzmann's constant, m denotes mass of gas particles, and T signifies the temperature of the gas. As temperature decreases, there's an increase in the wavelength (λ) of particles. This leads to overlapping wave functions of adjacent particles when the ratio of λ to the inter-particle distance (a) approaches unity. In realm of bosons, multiple particles experience condensation into state of lower energy.

The examined subject matter is frequently denoted as Bose-Einstein condensate (BEC). Particles occupying the identical energy level manifest a wave function that is mutually superimposed and possesses identical momentum and energy. The macroscopic wave function, represented by the symbol ψ, is a viable description of the particle system. At a temperature of 2.2°K, the condensation of bosons in 4He manifests superfluidity (Yoon & Cho, 2021).

2.2.2.3. Zero Resistance

The arrangement of electrons within a confined space is dictated by their fermionic nature, resulting in a configuration that follows an ascending order of energy levels, commencing from the lowest. As result, electron system's total energy is significantly high. System's total energy can decrease if a particular mechanism causes electrons to transform into bosons. BCS theory, introduced by Bardeen, Cooper, and Schrieffer in 1957, was based on this notion and provided a comprehensive explanation of the underlying principles of superconductivity. The BCS theory examines the properties of Cooper pairs, consisting of a pair of electrons that are paired together (Schwindt et al., 1999). Based on the preceding discussion, it can be deduced that Cooper's pairs demonstrate bosonic characteristics. As a result, the pairs experience a process of condensation towards a lower energy state when temperature is decreased to critical temperature (Tc). Electron system can be characterized by macroscopic wave function denoted as ψ, that arises from the superposition of multiple wave functions of Cooper pairs that exhibit identical wave functions. The variable ψ is defined in the following manner:

$$\Psi = \Psi_0 \exp(i\theta) \qquad (8)$$

Cooper pairs density is represented by the quantity indicated as $|\psi 0|^2$, whereas phase is denoted by θ. Macroscopic wave function, known as the order parameter ψ, demonstrates a specific form of stiffness recognized as rigidity. Furthermore, the energy linked to this occurrence is called the energy gap. A disruption of the Cooper pairs necessitates a value of 2Δ (Gatti et al., 2003).

The primary attribute of a superconductor is its ability to facilitate the flow of electrical current without experiencing any resistance. The phenomenon above pertains to the genesis of Cooper pairs and energy gap of 2Δ. During the state of normal conduction, the trajectory of an individual electron is susceptible to scattering by impurities and phonons, leading to a compelled alteration in its direction of motion. This phenomenon results in the production of electrical resistance. When in a state of superconductivity, the Cooper pairs exhibit uniform center of mass momentum (p), and the presence of a superconducting current is indicated by $p \neq 0$. In contrast, the current flow is nonexistent in cases where the value of p equals zero (Aspect & Grangier, 1987).

The presence of an electric current imposes limitations on the direction of motion of Cooper pairs, owing to their strong interaction. The previously mentioned phenomenon can solely transpire following the dissociation of a Cooper pair. The Cooper pairs must be endowed with kinetic energy proportional to the energy gap to attain the result above. This can be achieved by administering an electric current. The current's magnitude is substantial, indicating that the production of electrical resistance cannot be ascribed to a negligible current (Patra & Lal, 2021).

2.2.2.4. Origin of Cooper Pair Formation

The Coulombic repulsion, which arises from the electrostatic interaction between charged particles, is a phenomenon that manifests itself in the interactions between electrons. The energy associated with the system can be expressed mathematically as $V(r) = (1/4\pi\, \varepsilon_0)\, e^2/r$. Here, e denotes electrical charge, r represents distance between electrons, and ε_0 signifies vacuum permittivity. The formation of an electron pair necessitates the surmounting of the Coulombic interaction (Bohm et al., 2018). Coulombic interaction in a specific medium can be mathematically represented by:

$$\begin{aligned} V(r) &= (1/4\pi\varepsilon)e^2/r \\ &= (1/4\pi\varepsilon_0)e^2/r - (1/4\pi r)e^2(1/\varepsilon_0 - 1/\varepsilon), \end{aligned} \qquad (9)$$

The symbol ε is utilized to represent the permittivity of the given medium. The strength of the potential function V(r) may diminish if the permittivity of the medium, represented by ε, surpasses the permittivity of free space, represented by ε_0, where $\varepsilon_0 < \varepsilon$. The emergence of an attractive force between electrons that counteracts Coulombic repulsion is a resultant phenomenon. The phenomenon whereby an electron within a crystal lattice exerts an attractive force on adjacent positively charged ions, leading to the subsequent attraction of a second electron by same +ve charges, is observed. It suggests that an aesthetically pleasing interaction occurs within the medium between electrons, overcoming the barrier (Pal et al., 2022) (Figure 2.7).

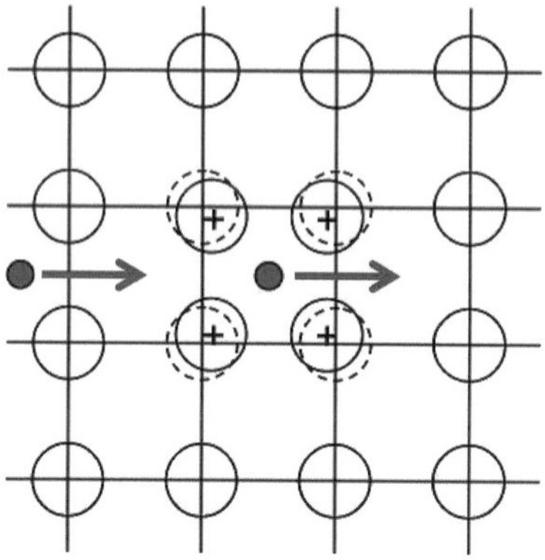

Figure 2.7. Phenomenon of electron-electron attractive force generation within crystal lattice. 2nd electron within crystal experiences an attractive force due to +ve charge of crystal ions. This positive charge arises from the Coulomb interaction with the first electron that traverses the same region. Hence, it can be observed that a pair of electrons engage in mutual interaction via an intermediary medium.

Source: Bhattacharya, Creative Commons License.

The Coulombic interaction, which operates at a proximate level, is implicated in the genesis of Cooper pairs. The phenomenon mentioned above is ascribed to the interplay between electrons and phonons, which provides a credible rationale for the genesis of traditional superconductors exhibiting low critical temperatures. The observation above provides evidence for creating Cooper pairs in superconductors with high-temperature properties. The precise mechanism underlying the attractive force manifested by a Cooper pair remains unresolved and is currently the focus of sustained inquiry (Gabovich et al., 2022).

The Cooper pair formation process necessitates considering the uncertainty principle, as denoted by Eq. (2). The correlation between Δ and kB Tc, specifically $\Delta \approx$ kB Tc, can be established due to Cooper pairs' disintegration at the critical temperature Tc. According to Equation 2, it can be inferred that the duration necessary for forming a Cooper pair is denoted by t_0 and is approximately equal to h/Δ. Approximations for t_0 at T_c = 100 K and Tc = 10 K were derived based on the provided relationship. The estimated values for t_0 were approximately 10^{-13} and 10^{-12}, respectively (Ruan et al., 2018).

It is significant to note that superconductivity manifestation depends on the time duration exceeding t_0. The BCS coherence length or Pippard coherence length is a commonly used term for the electron displacement denoted as ξ 0, which is necessary for pair formation. The quantity denoted as ξ 0 is $t_0 v_F$, with v_F signifying Fermi velocity. For instance, estimates of ξ 0 at distances of approximately 10^{-9} m and 10^{-6} m are computed utilizing typical Fermi velocities of roughly 104 m s^{-1} and 10^{6} m s^{-1}, respectively, while considering Tc values of 100 K and 10 K (Geerlings et al., 1990).

2.2.2.5. Josephson Effect

The present study introduces an additional implication that can be deduced from the expression of superconductivity via the macroscopic wave function. Given the scenario presented, it can be observed that 2 superconductors are connected using a thin insulating layer,(Bloch, 1968). Below are Schrödinger equations for superconductors 1 and 2.

$$i\hbar(\partial\psi_1/\partial t) = eV\psi_1 + K\psi_2,$$
$$i\hbar(\partial\psi_2/\partial t) = -eV\psi_2 + K\psi_1, \quad (9)$$

The symbol K represents coupling constant of wave functions across insulator, whereas the letter e denotes electronic charge. Furthermore, the symbol V denotes the magnitude of the electric potential difference imposed across the pair of superconducting materials (Jeanneret & Benz, 2009) (Figure 2.8).

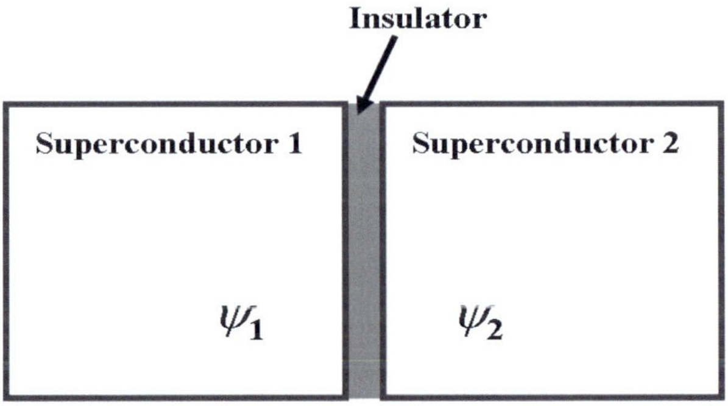

Figure 2.8. Josephson effect manifests in configuration wherein two superconducting entities are linked via a slender insulating stratum. The magnitude of the tunneling current Js in a superconducting system is contingent upon the disparity in phase between the macroscopic wave functions ψ_1 and ψ_2.

The Josephson equations can be obtained by utilizing the macroscopic wave functions of two superconductors, represented as ψ_1 and ψ_2. The distinctive features of each wave function are its corresponding super electron density, represented by n_1 and n_2, and phase, represented by θ_1 and

θ_2. By assuming the wave functions to be $\psi_1 = n_1^{12} \exp(i\theta_1)$ and $\psi_2 = n_2^{12}$ $\exp(i\theta_2)$ for each respective area, one can derive the Josephson equations.

$$J_s = J_c \sin\phi, \quad \partial\phi/\partial t = 2eV/\hbar, \qquad (11)$$

In superconductivity, the critical current is commonly denoted by the symbol Jc. In contrast, the phase difference between the first and second superconductors is represented by the symbol φ, where $\theta_2 - \theta_1$. The statement above suggests that superconducting tunneling current, represented as Js and pertains to phase differential, passes through junction. Furthermore, it is suggested that a junction can conduct an alternating current that is directly proportional to the applied voltage (V). The Josephson effect, as described by Josephson in 1962, showcases the expression of electron wave characteristics on a larger scale (Ambegaokar & Halperin, 1969).

2.2.3. Superconductivity and Thermodynamics

When superconductor is exposed to magnetic field, current produced demonstrates a complete lack of resistance, thereby setting it apart from the reaction observed in the state of normal conductivity. The study of quantized vortices is a topic of great interest, and manipulating vortex lines is of considerable significance for real-world applications (Linder & Sudbo, 2007).

2.2.4. Superconductor in Magnetic Field

The superconductor demonstrates the absence of resistance while simultaneously displaying magnetic flux density of B = 0. Meissner effect, is the phenomenon under discussion. Currently, the equation M = –H is derived from

$$B = \mu_0(H + M) = 0 \qquad (12)$$

In magnetism, the symbol H is commonly used to denote the magnetic field, while temperature represents by (T) and magnetic field (H), represents magnetization (Tomita & Murakami, 2003). The magnetic permeability within a vacuum is represented by the symbol μ 0. It is possible to formulate the superconductor's differential Gibbs free energy.

$$dG = -SdT - \mu_0 MdH \qquad (13)$$

The equation above describes the correlation between the thermodynamic property of entropy (S) and the physical quantity of temperature (T). At an isothermal condition T, the variation in Gibbs free energy (dG) can be represented as the inverse of the product of the magnetic moment (μ_0M) and the alteration in a magnetic field (dH). This can also be denoted as the product of the magnetic field ($\mu_0 H$) and the change in magnetic field (dH). By integrating Equation 1.13 within the range of H = 0 to H while in the superconducting state, it is possible to establish the disparity in energy (Pippard & Bragg, 1953).

$$G(T,H_s) - G_S(T, 0) = (1/2)\mu_0 H^2 \qquad (14)$$

The Gibbs free energy of a superconductor, represented as Gs(T, H), is obtained on a per-unit-volume basis of the superconducting substance. The application of a magnetic field to superconductor results in the disruption of its superconductivity state.

$$G = H_c(T) \qquad (15)$$

The symbol H_c (T) is used to represent the critical field. When no magnetization (M = 0) is caused by the application of H in normal conducting state, there's no observable change in energy. Thus, the Gibbs free energy of conventional conductor represented as G_n (T, H) = G_n (T, 0). Equation mentioned above is obtained through the superconductor's transition to normal conductor at H=H c (T), resulting in G_n (T, H_c) = G_s (T, H_c). Consequently, the discrepancy in the accessible energy is ascertained (Rosenstein & Li, 2010).

$$G_n(T, 0) - G_S(T, 0) = (1/2)\mu_0 H_c^2(T) \ (16)$$

The phenomenon of varying energy levels is frequently denoted as condensation energy in academic discourse.

2.2.4.1. London Equations

The Meissner effect is expounded by the fundamental equations presented in London theory. Motion of equation for superconducting electron, which demonstrates complete absence of resistance, can be mathematically represented as m multiplied by derivative of electron's velocity with respect to time, equaling negative charge e multiplied by electric field strength E. Additionally, superconducting current density formula, $\mathbf{J}_s = -n_S e * \mathbf{v}_s$, can be applied to derive the resulting equation (Lake et al., 2002) (Figure 2.9).

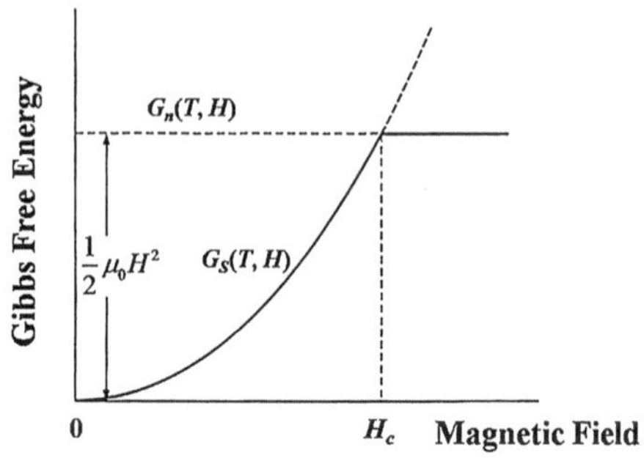

Figure 2.9. Gibbs free energies of normal and superconducting states depend on magnetic field. Eqn. (14) describes how the magnetic field affects Gibbs's free energy.

Source: P. Seidel, Creative Commons License.

$$\partial JS\partial = t(n\, e_s\, ^{*2}m^*)\mathbf{E} \quad (17)$$

The equation above takes into account the electric field (E), speed (vs), and density (ns) of a superconducting electron in addition to its mass (m) and charge (-e). It is significant to remember that the charge (e*) exceeds zero (e * > 0). The following holds in conjunction with Maxwell equations:

$$\nabla \times \mathbf{E} = -\partial B/\partial t \quad (18)$$

The resulting expression is obtained Upon substituting the magnetic flux density variable B (Brandt & Indenbom, 1993).

$$\frac{\partial}{\partial t}\left(\frac{n_s e^{*2}}{m^*} B + \nabla \times J_s\right) = 0 \quad (19)$$

Assuming that expression within parentheses is equivalent to zero, one can obtain subsequent equation through the following derivation.

$$\frac{n_s e^{*2}}{m^*} B + \nabla \times J_s = 0 \quad (20)$$

A request has been made for an elucidation of Meissner effect. Equations (17) and (20) are commonly called 1st and 2nd London equations. Once again, in collaboration with.

$$\nabla \times = \mathbf{B}\, \mu_0 \mathbf{J}_s \quad (21)$$

we have:

$$\nabla^2 \mathbf{B} = (1/\lambda_L)^2 \mathbf{B} \quad (22)$$

The value of London penetration depth, denoted as λL, can be expressed as follows:

$\lambda_L = (m^*/n\ e_s{}^{*2}\mu_0)^{1/2}$ (23)

Equations (22) and (23) can be utilized to determine the surface magnetic field of the superconductor. The signal represented by the symbol B, undergoes a reduction in amplitude due to the presence of surface within range of wavelength λ L. Magnetic flux density B formulation can be expressed (Albash & Johnson, 2008) (Figure 2.10).

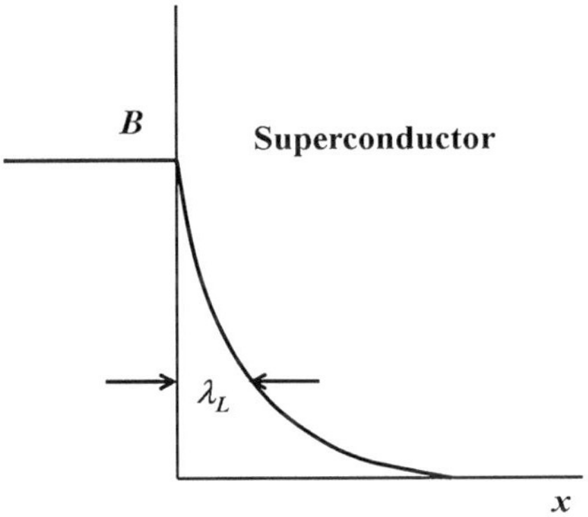

Figure 2.10. The diagram presented depicts diminution of magnetic flux density B at surface of superconductor. The Meissner effect is observed in superconductors that have thickness exceeding range of λ L, despite the fact that magnetic field is capable of permeating the superconductor within this range.

Source: Joachim Albrecht, Creative Commons License.

$\mathbf{B} = \nabla \times \mathbf{A}$ (24)

The symbol A denotes the magnetic vector potential. It is worth noting that following important equation is presented.

$J = -(n\ es\ {}^*2\ m^*)\mathbf{A}$ (25)

Equations (20) and (24) employment is viable, given the appropriate conditions. The Meissner effect is established in a superconductor when a magnetic field is applied, inducing a stationary supercurrent through the influence of the vector potential.

The values of $e^* = 2$ e (denoting double electronic charge) as 3.2×10^{-19} C, $m^* = 2$ m (denoting double electron mass) as 11.8×10^{-30} kg, and $n_s = 5 \times 10^{27}$ m^{-3}, can be utilized to approximate λ L, which is approximately 53 nm. Notably, the value of λ L is approximately 50 nm in aluminum and approximately 200 nm in niobium. The Meissner effect is an observable occurrence in superconductors that possess a thickness exceeding penetration depth (λ L). In such cases, the superconducting current that flows to the area within the λ L depth from the surface effectively repels the external magnetic field (Edwards, 1981).

2.2.4.2. Ginzburg-Landau Theory

The London equation does not account for coherence length ξ within its formulation. In 1950, Ginzburg and Landau introduced phenomenological theory of the superconductor, which is well-documented in their publication. The employed theory incorporates the employment of an order parameter denoted as $\psi = |\psi_0|\exp(i\theta)$. As per the Ginzburg-Landau (GL) theory, the Helmholtz free energy per unit volume of the superconductor can be expressed as a Taylor series of the modulus squared of the order parameter (|ψ|2), subject to the condition that the magnitude of ψ is adequately small (London & London, 1935). Therefore, it can be deduced that we have.

$$F_s = F_n + \alpha|\psi|^2 + (\beta/2)|\psi|^4 + \cdots, \quad (26)$$

The Helmholtz free energies corresponding to superconducting and normal conducting states are Fs and Fn, respectively. The symbols represent the coefficients of expansion α and β (Balachandran & Teotonio, 1993) (Figure 2.11).

$$\alpha()T > 0, \beta > 0\ (T > T_{c)}\ \alpha()T < 0, \beta > 0\ (T < T_c)(27)$$

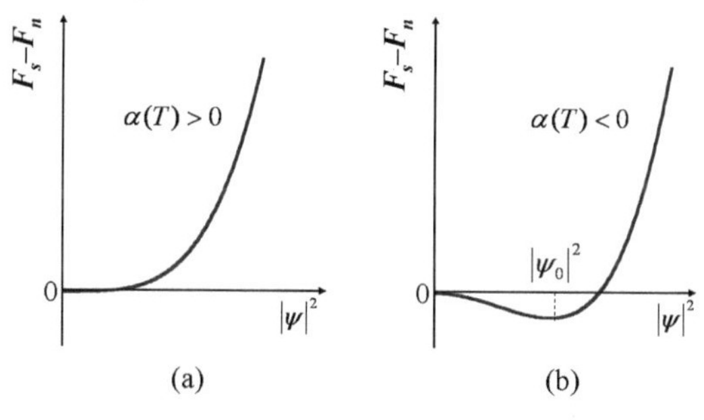

Figure 2.11. The present study concerns the Ginzburg-Landau free energy functions denoted as F_s–F_n, which are analyzed for two distinct temperature regimes: (a) T > Tc (α > 0); and (b) T < Tc (α < 0). The Helmholtz free energies corresponding to the superconducting and normal conducting states are represented by the symbols F_s and F_n, respectively.

Source: Jonas Bekaert, Creative Commons License.

That is, in $T < T_c$, $|\psi| \neq 0$ exists, and

$$|\psi| = |\psi_0| = (\alpha/\beta)^{1/2} = (\alpha_0/\beta)^{1/2}(T_c - T). \quad (28)$$

The statement mentioned above suggests that the superconducting state displays a state of constancy either at or beneath the critical temperature, symbolized as Tc. Moreover, it is possible to derive a correlation between the superconducting electron density, denoted as ns (r), and the absolute value squared of the wave function, represented as $|\psi(r)|^2 = ns(r)$.

The next equation gives the free energy in the magnetic fi eld:

$$F_s = F_n + \alpha|\psi|^2 + (\beta/2)|\psi|^4 + (1/2m^*)|(-i\hbar\nabla + e^*\mathbf{A})\psi|^2 + (1/2\mu_0)\mathbf{B}^2. \quad (29)$$

The following formulas are derived from the free energy using variation derivation with regard to and A (Duan et al., 1998).

$$\alpha\psi + \beta|\psi|^2\psi + (1/2m^*)(-i\hbar\nabla + e^*\mathbf{A})^2\psi = 0, \quad (30)$$

$$\mathrm{J}_s = (e^*\hbar/2m^*i)(\psi^*\nabla\psi - \psi\nabla\psi^*) - (e^{*2}/m^*)|\psi|^2\mathbf{A}. \quad (31)$$

GL theory incorporates results of London theory, as Equation 31, which fulfills the requirements of $\nabla\psi = 0$ and $\nabla\psi^* = 0$, is equivalent to Equation 25. In the one-dimensional system, it is postulated that the wave function ψ has a small magnitude and that the potential energy A is zero, leading to the ensuing result.

$$(\hbar^2/2m^*)(\partial^2\psi/\partial x^2) = \alpha\psi \quad (32)$$

$$\psi = |\psi_0|\exp(-x/\xi), \quad (33)$$

ψ in this equation is:

$$\xi(T) = (\hbar^2/2m^*\alpha(T))^{1/2}. \quad (34)$$

The Ginzburg-Landau theory's coherence length pertains to the distance scale at which the order parameter can effectively recuperate from spatial fluctuations. In contrast, the coherence length ξ 0, also referred to as the Pippard or BCS coherence length, can be obtained through the uncertainty principle and denotes the typical length that signifies the extent of the Cooper pair (Ross, 1983).

2.2.4.3. *Type I and Type II Superconductors*

The GL theory encompasses a dual degree of freedom system, specifically the phase and amplitude-integrated into the order parameter. The GL coherence length ξ(T) and the GL penetration depth λ(T) are representative lengths demonstrating spatial correlations of the amplitude and the phase. The lengths mentioned earlier depend on temperature and display variability based on the expression $(1 - T/T\,c)^{-1/2}$ in relation to temperature. GL parameter is a fundamental factor that governs superconductor magnetic characteristics, which is determined by the ratio between the temperature-dependent London penetration depth and the coherence length. An investigation into the effect of the parameter κ on the state of superconductivity is necessary, where κ is defined as the ratio of the temperature-dependent London penetration depth λ(T) to the temperature-dependent coherence length ξ(T) (Liu, 1998).

When a superconductor is invaded by a localized external magnetic field, the superconductivity within the area of radius ξ is disturbed, leading to its loss. Simultaneously, the energy of the magnetic field experiences a decrease within the region of radius λ, where there is an absence of magnetization. The phenomenon above results in an elevation of the free energy of Gcore by a factor of $(1\ 2)\mu\ \pi\xi_0{}^2Hc^2$ per unit length (Goodman, 1964). $G_{mag} = (1/2)\mu\ \pi_0 H_e{}^2\lambda^2$ per unit length is decreased. The total free energy can be obtained as follows:

$$G = G_{core} + G_{mag} = (1\ /2)\mu\ \pi\ \xi_0 H_c{}^{22} - (1\ /2)\mu\ \pi_0 H_e{}^2\lambda^2. \quad (35)$$

As per Equation 35, it can be hypothesized that the favored result is the deterioration of a fraction of the superconductive area when $G < 0$. On other hand, presence of a magnetic field is inhibited when value of G is greater than zero. Lower critical field H_{c1} can be obtained by setting H_e equal to H_{c1} and postulating G to be 0, where Hc is critical field and ξ and λ are coherence length and penetration depth, respectively. The value of G likely is less than zero when the value of H surpasses H_{c1} at a κ greater than one. Drawing upon the GL theory, a meticulous examination elucidates that the boundary between G values less than zero and those greater than zero can be articulated as $\kappa = 1\ \sqrt{2}$, yielding $H_{c1} = (Hc\ 2\kappa)\ln\kappa$ (Cyrot, 1973). Therefore, the following criteria can be obtained.

- $\kappa < 1\sqrt{2}$ $H < H_c$ Meissner state:
  $H > H_c$ Normal conducting state
- $\kappa > 1\sqrt{2}$ $H < H_{c1}$ Meissner state:
  $H > H_{c1}$ Magnetic fi eld invades partially.

Electrical resistance disappears entirely in presence of a magnetic field in type I superconductors, while type II superconductors exhibit zero electrical resistance while permitting the penetration of magnetic fields (Milosevic & Geurts, 2010).

2.2.4.4. *Mixed State*

Various vortices can be observed within the natural domain. Induction of a magnetic field onto type II superconductor leads to multiple vortices within the superconductive substance. This phenomenon is elucidated as follows (Babaev & Speight, 2005).

The mixed state in type II superconductor is distinguished by the penetration of a magnetic field in the shape of multiple vortices, which arise when magnetic field strength exceeds lower critical field, H_{c1}. Vortex's complex cross-sectional structure. The radius of the magnetic field encompassing a single vortex line in type II superconductor is represented by λ, and it surpasses radius of the normal conducting core, which is denoted as ξ. (Ling et al., 2013) (Figure 2.12).

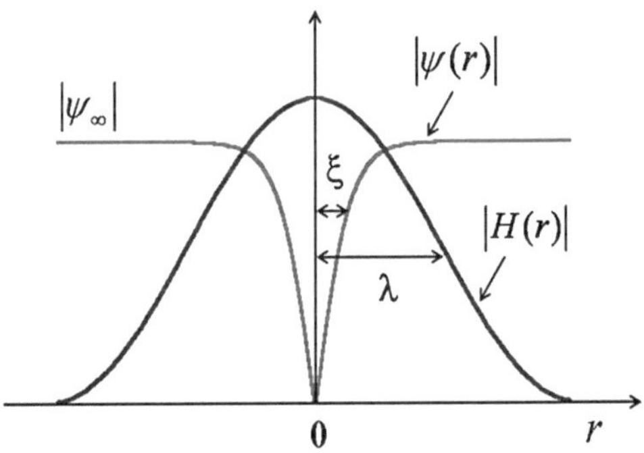

Figure 2.12. Present illustration depicts the schematic diagram of vortex structure that occurs in type II superconductor. Vortex comprises core of normal conductivity, characterized by a radius of ξ, enveloped by magnetic field of radius λ. Quantization of confined magnetic flux is an established phenomenon, with a specific value of Φ 0 = 2.07 × 10^{-15} Wb.

Source: Inna Vishik, Creative Commons License.

This particular arrangement is commonly denoted as the flux line in academic discourse. The magnetic flux is confined to a specific region characterized by the wavelength λ due to the circulation of a current that shields the magnetic field. This current exhibits zero resistance. 47 The quantization of magnetic flux under confinement is commonly called "fluxon." The specific value of this quantized magnetic flux is represented by the symbol Φ 0, which is equal to 2.07× 10–15 Wb. Occurrence of shielding current is distinguished by a macroscopic wave function represented as ψ. It is important to note that the satisfaction of Bohr's quantization condition depends on shielding current's completion of a circuit around the vortex. Aforementioned refers to phenomenon of quantization of magnetic flux. It is important to acknowledge that term "flux line" applies solely to conventional type II superconductors. Characterization of flux lines in high-Tc cuprates, classified as extreme type II superconductors, presents a challenge due to the elevated λ to ξ ratio. Therefore, the notion of a vortex line holds importance (Chen et al., 2021).

The Meissner state with a magnetization (M) equal to negative applied magnetic field (H) can be visually detected in the M-H curve. This phenomenon occurs within the $0 < H < H_{c1}$ range. In mixed phase of superconductor, presence of an external magnetic field with a strength greater than the lower critical field ($H > H_{c1}$) allows for the maintenance of superconductivity until upper critical field ($H = H_{c2}$) is attained. At this juncture, the entire region of the superconductor becomes inundated with vortices. The notation H_{c2} is commonly used to represent the upper critical field. Critical magnetic field values, represented as $\mu_0 H_{c2}$ ($=B_{c2}$), exhibit a typical range of 10–30 T for metallic superconductors. On the other hand, it is noteworthy that high-temperature superconductors demonstrate critical magnetic field values that exceed 100 T. Type II superconductors are commonly employed in practical applications owing to their elevated $\mu_0 H_{c2}$ values (Zampoli & Jordan, 2022). The mathematical formula representing the upper critical field can be expressed as follows:

$$\mu 0Hc2 = Bc2 = \Phi 0\ 2\pi\xi^2 \quad (36)$$

The emergence of triangular lattices of vortices is observed in a superconductor in its original state and possessing no pinning centers. The Abrikosov lattice, vortex lattice, or flux line lattice are all common names for the triangular lattice. The triangular vortex lattice's lattice interval, represented by the symbol f, can be mathematically expressed in the following manner:

$$a_f = (2\Phi_0\ \sqrt{3}B)^{1/2} \quad (37)$$

What is the effect of applying a current density J to a superconductor in the mixed state?

Currently, the Lorentz force is being exerted by the electric and magnetic fields:

$$\mathbf{f}_L = \mathbf{J} \times \mathbf{e}_z \Phi_0 \quad (38)$$

On the vortices here, $\mathbf{e}_z$ is the unit vector in the z-direction. Moving the vortices at a speed of $\mathbf{v}_L$ by this force, the electric field is generated.

$$\mathbf{E} = \mathbf{B} \times \mathbf{v}_L \quad (39)$$

The occurrence distinguished by the displacement of vortices is frequently denoted as either vortex flow or flux flow. The motion of vortices is subject to frictional forces from their surrounding environment, which restricts their velocity and is commonly represented as v L. The opposing force of friction is offset by the viscous force, denoted by $-\eta$ v L, where the viscosity coefficient is represented by η. Consequently, variable L does not demonstrate unbounded growth. The practical implementation of superconductivity is significantly challenged by energy dissipation, particularly in the vortex flow state (Aston et al., 1971).

2.3. SUPERCONDUCTING MATERIALS

Superconductivity has been observed in different materials, including approximately 2,000 types of elements, alloys, diverse metallic compounds, and organic substances. The phenomenon of superconductivity is widely observed. The subsequent discussion elaborates on the current state of our comprehension regarding traditional superconducting substances (Norman et al., 1995).

Remember

Superconducting compounds, including organic superconductors, offer diverse opportunities for exploring and enhancing the properties of high-TC superconductors.

2.3.1. Superconducting Metals and Alloys

2.3.1.1. Elements

After discovery of superconductivity in mercury (Hg) in 1911, subsequent research confirmed the possibility of lead (Pb) and tin (Sn) exhibiting superconductivity. The process of identifying superconducting elements was subsequently carried out gradually and systematically. Above 30 elements demonstrate the phenomenon of superconductivity when subjected to extremely low temperatures. Moreover, it has been confirmed that over 20 supplementary elements demonstrate superconductivity when subjected to high pressure. Around half of the elements listed in periodic table are currently acknowledged as potential superconductors. Nevertheless, verification of superconductivity in alkali metals, specifically Na, magnetic materials like Co and Ni, materials with low resistivity including Au, Ag and Cu, remains outstanding (Lowell & Sousa, 1970). The only superconducting elements considered viable are Pb and Nb, as they demonstrate Tc values that exceed liquid helium temperature(4.2 K). Niobium has a body-centered cubic (bcc) crystal lattice and critical temperature (Tc) of 9.25°K. The substance referred to as Nb demonstrates attributes of a superconductor of type II, with an associated magnitude of roughly 0.3 T for its $\mu_0 H_{c2}$. One can derive the relationship between resistivity (ρ) and the square of the magnetic field strength ($\mu_0 H_{c2}$) as follows.

$$\mu_0 H_{c2} = B_{c2} = 3.1 \times 10^3 \mu \rho \gamma_0 T_c \qquad (40)$$

The symbol denotes the electronic specific heat coefficient γ. Including oxygen or nitrogen as impurities in Nb leads to an elevation in $\mu 0H_{c2}$, owing to the simultaneous rise in ρ. The thermal conductivity of a material may undergo modifications due to the existence of impurities. Typically, the diminution of Tc is frequently observed following the integration of oxygen or nitrogen atoms into the crystal lattice (Scanlan et al., 2004) (Figure 2.13).

1	2	3	4	5	6	7	8	9	10	11	12	13	14	15	16	17	18
Li 20	Be 0.03											B 6.0	C 4.0		O 0.6		
												Al 1.19	Si 6.7	P 4.6 6.1	S 17.0		
		Sc 0.05	Ti 0.39	V 5.3			Fe 2.0				Zn 0.9	Ga 1.09	Ge 5.4	As 0.5	Se 6.9		
		Y 1.5 2.7	Zr 0.55	Nb 9.2	Mo 0.92	Tc 7.8	Ru 0.5	Rh 3.2×10^{-4}			Cd 0.55	In 3.4	Sn 3.7 5.3	Sb 3.6	Te 4.5	I 1.2	
Cs 1.5	Ba 1.8 5.1	La 4.9α 6.3β	Hf 0.13	Ta 4.4	W 0.01	Re 1.7	Os 0.65	Ir 0.14	Pt 0.002		Hg 4.15 3.95	Tl 2.39 1.45	Pb 7.2	Bi 3.9 7.2 8.5			

Ce 1.7						Gd 1.08							Lu 0.1 ~0.7
Th 1.37	Pa 1.3	U 0.2	Np 0.07		Am 0.8								

Figure 2.13. T_c of the periodic table's superconducting elements. The T_c value is displayed in the box, while shaded boxes indicate elements that exhibit superconductivity under high pressure.

Source: G. Zwicknagl, Creative Commons License.

Did You Know?

Cuprate superconductors, a type of high-TC oxide superconductor, have unique layered crystal structures that play a significant role in their superconducting behavior.

The electron density of the N(EF) state within the Fermi surface is modified by the presence of the impurity atom. On the contrary, the Pb crystal manifests a face-centered cubic (fcc) configuration, and its crucial temperature (Tc) is prone to alteration upon the inclusion of minute amounts of dopant impurities. Introducing doping agents, such as Hg or Tl, leads to a considerable decrease in Tc. This phenomenon can be observed as an instance. On the other hand, introducing Bi, As, or Sn through doping results in a rise in Tc of roughly 1 K. The element Pb, initially categorized as type I superconductor, experiences a transition to type II superconductor due to an increase in ρ caused by addition of small amounts of doping impurities (Yao & Ma, 2021).

2.3.1.2. Alloys

The quantity of superconducting alloys surpasses one thousand. The composite material displays a crystalline structure and physical characteristics that are akin to those of bulk metallic products, with gradual modifications occurring upon the incorporation of minor constituents. Nevertheless, the superconducting characteristics of metals surpass those of pure metals. The equilibrium diagram of Nb-Ti alloy, a material of significant importance in the field of superconductivity and extensively utilized in various industrial applications.(Hirsch et al., 2015).

The presence of the β phase in Nb-Ti alloys at low temperatures can be attributed to the mass-centered cubic crystal structure of both Nb and Ti. The outcome of this procedure yields a composite comprising of niobium and titanium within the temperature spectrum of the illustrations. At temperatures equal to or below 883°C, the transformation of Ti results in the formation of a hexagonal tight-packed cage (hcp). Subsequently, when the temperature falls below this particular threshold, the α-Ti and β'-NbTi phases amalgamate with the preceding fraction of Nb that scatters into Ti. The critical temperature (Tc) can be mathematically expressed based on the weak bonding theory derived from BCS (Hott et al., 2013).

$$T_c = 1.13\theta_D \exp(-1/N(E_F)V) \quad (41)$$

The above formula integrates the Debye temperature, denoted by the symbol θ D, the density of electron states at the Fermi surface, represented by N (E F), and an electron interaction component, symbolized as V. The β-phase demonstrates peak values of N (E F), V, and γ in the vicinity of 40 at %Ti, resulting in a decrease in θ D to its minimum value. As a result, the maximum critical temperature (Tc) of this region is achieved. Conversely, it has been previously elucidated that the value of $\mu_0 H_{c2}$ is correlated with Tc, γ, and ρ, and it reaches its maximum level in the vicinity of 50% Ti. Although the Nb-Zr alloy was developed before the Nb-Ti alloy, it is no longer in use due to its lower $\mu_0 H_{c2}$ value compared to Nb-Ti and the challenges associated with its processing (Larbalestier et al., 2001).

2.3.2. Superconducting Compounds

2.3.2.1. A 15 Type of Compound

The A_{15}-type compound superconductors hold significant importance as they are deemed suitable for practical applications. The aforementioned compounds, specifically Nb_3Ge, Nb_3Sn, and Nb_3Al, pertain to the A_{15} family and are noteworthy for their unique properties. The A_{15} compounds are known for their superconducting properties, with Nb_3Ge being notable for having the highest critical temperature (Tc) among them. On the other hand, Nb_3Sn is widely employed due to its favorable characteristics. Finally, Nb_3Al has been identified as the most recent member of the A_{15} group (Nakamura et al., 2007).

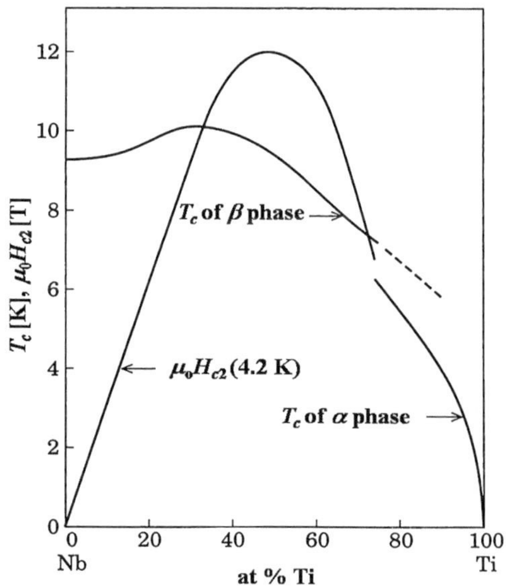

Figure 2.14. Graphical representation of T_c and $0H_{c2}$ variations in the Nb–Ti alloy as a function of Ti concentration. About 47 wt.% Ti constitutes the optimal Ti concentration.

Source: Koname, Creative Commons License.

The category of A_{15} compounds demonstrates a crystalline structure that is cubic in nature. The crystal architecture denoted by the chemical formula A_3B, as exemplified by Nb_3Sn, is composed of two discrete arrays of atoms, namely A and B. (Abrikosov, 1957) (Figure 2.14).

The critical temperature (Tc) and upper critical magnetic field (μ $0H_{c2}$) of A_{15} materials exhibit higher values compared to those of Nb alloys. The critical temperature (Tc) of the material under consideration reaches its maximum value when the atomic ratio of A to B is stoichiometric, which is in contrast to the behavior observed in the alloy system. The A atoms are positioned orthogonally to one another on the orthogonalization side of the bcc lattice, resembling a chain-like structure (Abrikosov & Gorkov, 1960). In contrast, the B atoms display a body-centered cubic (bcc) arrangement. The A atom next to it in the chain is its closest neighbor. The development of a pseudo-conducting channel by the mutual adhesion of A atoms, which results in a strong metallic link that is unidirectional, is what distinguishes the A_{15} crystal type. The persistence of the A-A atomic chain is responsible for the improved superconducting characteristics of materials with the formula A_{15} (Maki, 1964).

The stoichiometric ratio of the A_{15}-type compound is not invariably 3:1. The compositions of Nb_3, Al, and Nb_3G exhibit a higher concentration of Nb than anticipated by their respective formulas at low processing temperatures, despite displaying a 3:1 composition at high processing temperatures. The composition of Nb_3Ge is not in a 3:1 ratio when subjected to processing at either high or low temperatures. Furthermore, the A_{15} crystal lattice cannot be synthesized spontaneously through thin-film deposition via vapor phase or under standard temperature and pressure conditions, as it is not present in the Nb_3Si compound under such circumstances (Usadel, 1970).

The Tc values of no less than fifty superconductors of the A_{15} type have been duly confirmed. The attainment of higher Tc values necessitates the presence of a stoichiometric composition and a crystal structure that is characterized by a high degree of uniformity. The attainment of the aforementioned ideal parameters directly correlates with an increase in Tc. It is possible to expeditiously produce a high-quality crystal in a state of equilibrium using the stoichiometric compositions of Nb_3Sn, V_3Si, and V_3Ga. Thus, the utilization of these compounds for production of superconducting wires has already made significant progress (Abrikosov & Gorkov, 1958). Production of superconducting wires poses a challenge in the case of Nb_3 Al, Nb_3Ge, and Nb $_3Si$, due to the significant differences between their equilibrium compositions and stoichiometric compositions. In these instances, the production of a high-Tc crystal is unattainable unless it is synthesized under non-equilibrium circumstances.

Additionally, atomic order irregularity has significant impact on Tc. Because A atoms form pseudo-conducting chains, substituting B atoms results in the scission of the conducting channel. Therefore, obtaining a perfect 3:1 composition of the A_{15} phase does not improve Tc (Skalski et al., 1964).

2.3.2.2. Nitride and Chevrel Compound

The superconductor NbN possesses a NaCl crystal structure and exhibits a critical temperature (Tc) exceeding 17K, as reported in reference. Compounds such as TaN, NbC, and MoC exhibit superconducting transition temperatures (Tc) that surpass 10°K (Matthias et al., 1961).

The high T_c exhibited by NbN renders it a material of significant practical interest. The material in question exhibits a relatively low N (E_F) value, and it is hypothesized that the strong interaction between electrons and phonons is responsible for its elevated T_c. Typically, the fabrication of NbN involves utilizing the vapor deposition technique, producing microcrystalline films that demonstrate a $\mu_0 H_{c2}$ value of 44 T. In contrast, the bulk material only exhibits a value of 8T (Ekin, 1984).

The chemical formula for the Chevrel compound is $M_xMo_6S_8$, where M denotes a variety of metals such as Pb, Sn, Cu, rare-earth metals, or other metals. Chevrel compound exhibits $\mu 0H_{c2}$ value that surpasses 60T, despite having a lower T_c than an A_{15}-type compound. The compound Mo_6S_8 exhibits a cluster formation, whereby eight clusters are weakly united through a Pb atom. Interatomic distance among Mo atoms within cluster is comparatively shorter (ranging from 0.26 to 0.28 nm) compared to the interatomic distance among the Mo atoms outside the cluster (measuring 0.387 nm) (Matthias et al., 1954). Consequently, the electrons responsible for conduction are confined to the cluster and exhibit a distinctive electronic configuration. The $PbMo_6S_8$ compound exhibits mean free path of electrons at level of lattice constant, which is approximately 1/7 of that observed in Nb_3Sn. This significantly increases resistivity, resulting in a large value of $\mu_0 H_{c2}$ as per Equation 1.40. According to current understanding, electronic coupling among clusters is comparatively less potent than within a cluster. As a result, $PbMo_6S_8$ may be regarded as a quasi-zero-dimensional substance (Hor et al., 1987).

2.3.2.3. MgB_2

For a considerable duration, there was no enhancement in the maximum Tc exhibited by the metallic superconductors belonging to the A_{15} category. The discovery of the MgB_2 superconductor in 2001 had a significant impact on this scenario. The critical temperature (Tc) of the aforementioned material is 39°K, which currently stands as the most elevated value discovered for a metallic superconductor. The critical temperature (Tc) can be ascertained through the utilization of three parameters, namely θ D, N (EF), and V, as per the BCS theory as represented by Equation 1.41 (Sweedler et al., 1978). Preliminary examination of isotope effect revealed temperature modification of 1°K upon the replacement of boron 10B atoms with 11B atoms in MgB_2. This substitution confirmed the presence of a robust electron-phonon interaction with the B atom. The high critical temperature (Tc) in MgB_2 is believed to be caused by the presence of large V, as opposed to a particularly high N (EF) value, which is observed in materials such as Nb_3Sn (Ubaldini et al., 2010).

The structural configuration of MgB_2 bears resemblance to that of a hexagonal close-packed (hcp) arrangement. The B atoms arrange themselves in a honeycomb structure which is intercalated triangular lattice formed by Mg atoms. These two structures are then vertically stacked along c-axis of MgB_2 material. Magnesium (Mg) is a divalent element that undergoes ionization and provides a single electron per boron (B) atom. Honeycomb configuration observed in MgB_2, which is composed of B atoms, bears a striking resemblance to that of benzene (Hamilton et al., 1965). However, it is important to note that the primary distinction between the two is that honeycomb lattice comprised of B atoms constitutes a two-dimensional framework, which allows for the free movement of electrons throughout network. The π orbital band present in MgB_2 is responsible for facilitating conduction within the B layer. Furthermore, it facilitates conduction along the c-axis orientation by connecting the upper and lower B layers via inactive Mg ions. Conversely,mobility of electrons within σ band is constrained within B layer, thereby limiting their contribution solely to the conduction within said layer. MgB_2 exhibits the presence of two distinct bands that exist in disparate dimensions (Tarascon et al., 1988).

According to existing literature, MgB_2 is considered to be a superconductor with multiple energy gaps, specifically two, which is attributed to the varying electron-phonon coupling constants between the σ and π bands. The ongoing investigation pertaining to the production of wires and thin films composed of MgB_2 is making progress. The thin films of MgB_2 exhibit a remarkably high resistivity, leading to a significant increase in $\mu 0H_{c2}$ up to 51 T. An analogous pattern is discernible in thin layers of NbN (Saito & Yoshida, 2011).

2.3.3. Organic Superconductors

The initial production of an organic superconductor utilizing organic compound TMTSF dates back to 1980. Subsequently, the synthesis of (BEDT-TTF)2X superconductor, which boasts a Tc exceeding 10 K, was achieved. TMTSF functions as an electron donor, and its combination with ClO_4 acceptors results in the formation of TMTSF molecules arranged in a pillar-like configuration. One-dimensional conduction is observed due to the overlapping of π electrons in a specific direction. The BEDT-TTF molecule is known to function as a donor in certain chemical reactions (Urayama et al., 1988). The π electrons are present on network crystal face comprising sulfur (S) in conjunction with iodine (I) acceptors, leading to two-dimensional conduction. The materials in question exhibit behavior similar to that of BCS theory, and experimental evidence has confirmed a critical temperature (Tc) exceeding 10 K in β (BEDT-TTF)-$2ICl_2$ superconductor when subjected to high pressure (Tachiki & Takahashi, 1989) (Figure 2.15).

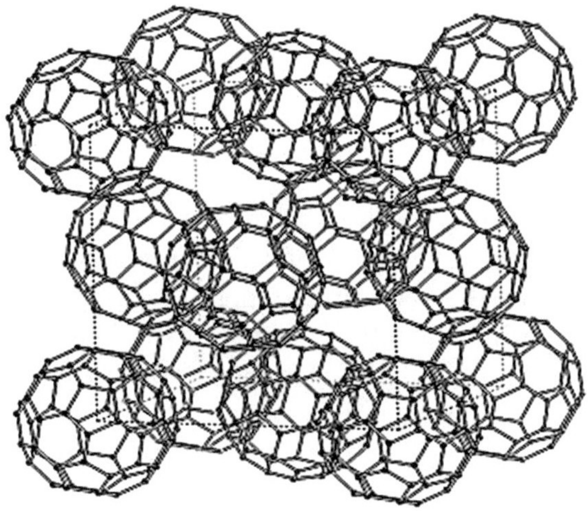

Figure 2.15. C_{60} molecule-based fcc structure. C_{60} is not a superconductor by itself, but it turns into one when alkali metals are added.

Source: Pubs, Creative Commons License.

The metallic atoms act as charge carriers and are located within this region. Superconductivity is achieved through the flow of carriers within adjacent C_{60} molecules. The initial C_{60} superconductor identified was K_3C_{60}, which exhibited a T_c value of 18K. Subsequently, Cs_3C_{60} was discovered to possess a significantly higher T_c value of 38K. The superconducting transition temperature (T_c) escalates with increased lattice constant. This can be attributed to the reduction in π orbital overlap of C_{60} as lattice spacing extends, leading to a narrower bandwidth and consequent increase in density of states at the Fermi level (N (E_F)) (Maeda et al., 1988).

The superconducting materials in question exhibit a notably elevated μ_0H_{c2}, with K_3C_{60} and Rb_3C_{60} achieving 54 T and 78 T values, respectively. Isotope effect observation indicates that these materials exhibit characteristics of BCS superconductors, with significant interaction between electrons and phonons (Schuller & Jorgensen, 1989).

2.4. HIGH-T_C OXIDE SUPERCONDUCTORS

A vast array of oxide materials currently exists, and it is noteworthy that the oxide superconductor currently holds the highest T_c. Therefore, the focus of this book centers on oxide superconductors, which encompasses the highest-T_c superconductor. This text will provide a concise overview of these materials (Murakami et al., 1990).

2.4.1. Early Oxide Superconductors

Before high-temperature cuprate superconductors discovery, numerous oxide superconductors were subject to investigation. $LiTi_2O_4$, which has a critical temperature (Tc) of approximately 14°K, along with $SrTiO_{3-x}$ (with Tc of approximately 0.5°K) and $BaPb_{1-x}Bi_xO_3$ (with a Tc of approximately 13°K) have generated significant interest due to their unique electron-phonon interaction (Ogata & Fukuyama, 2008).

Host crystal structure of $BaPb_{1-x}Bi_xO_3$ and $Ba_{1-x}K_xBiO_3$, the latter discovered subsequently, is $BaBiO_3$. The structural configuration commonly referred to as perovskite structure is denoted by ABO_3 chemical formula, wherein A and B denote metallic atoms. The perovskite structure, comprises atom A at origin and atom B at body-centered position. Oxygen atoms encompassing the B atom arrange themselves in an octahedral configuration, thereby establishing a three-dimensional network of BO_6 octahedra (Tsuei & Kirtley, 2000). The Bi atom at B site in ABO_3 structure exhibits a formal electric charge of Bi^{4+} (=$6s^1$) in the compound $BaBiO_3$. When in isolation, the Bi element exhibits metallic properties due to the presence of an electron occupying half of the 6s band. The robust electron-phonon interaction results in a significant coupling between electric charge of Bi atom and "breezing" mode, that corresponds to oscillatory expansion and contraction of the BiO_6 octahedron. Consequently, periodic lattice distortion and electronic density concentration give rise to a charge-density wave state wherein electric charge transitions from Bi^{3+} to Bi^{5+} at each site, ultimately leading to an insulating effect (Park & Synder, 1995).

Replacing Bi site with a Pb ion diminishes this strong electron-proton interaction, and metallic state is observed in example of $BaPb_{1-x}Bi_xO_3$. In other words, Pb $^{4+}$ (=6s ^{0}) dilutes Bi $^{4+}$ (=6s ^{1}), and average formal valence and filling of 6 s electronic band are thought to be reduced. $BaPb_{1-x}Bi_xO_3$ superconductivity is close to the metal-insulator transition barrier (Norman & Pepin 2003).

Like high-T_c cuprate superconductors, metals or superconductors can be produced by altering the band's filling factor in insulating materials due to strong electron correlation. The conduction network will become disorganized if the Pb site in $BaPb_{1-x}Bi_xO_3$ is substituted for the Bi site. Investigating the $Ba_{1-x}K_xBiO_3$ system, where K + ion replaces the Ba 2+ site that isn't directly involved in conduction, revealed $Ba_{1-x}K_xBiO_3$ metalizes when x.4 and simultaneously becomes superconductor with 0 $T_c \approx 30$ K. Until 2008, this was oxide with greatest T_c (apart from cuprate superconductors) (Damascelli et al., 2003).

2.4.2. Cuprate Superconductors

2.4.2.1. Crystal Structure

Copper oxide superconductor known as $La_{2-x}Ba_xCuO_4$ was first discovered in 1986. This substance was the precursor to the subsequent array of high-temperature superconductor findings. The fundamental component of high-temperature superconductors is CuO_2 layer. Stratum serves as the foundation for the manifestation of superconductivity. (Ando et al., 2004).

The K_2NiF_4 structure is present in $La_{2-x}Ba_xCuO_4$, wherein Ba $^{2+}$ ions substitute some of La 3+ ions. The material in question exhibits a layered structure wherein (La, Ba) layers and CuO_6 octahedrons are arranged alternatingly, while CuO_2 layers are oriented parallel to the bottom surface. The antiferromagnetic state is inherent to $LaCuO_4$. However, introducing Ba^{2+} ions at increasing concentrations results in the implantation of carrier holes into the CuO_2 layers. This process is accompanied by a concurrent increase in Tc, which reaches maximum value of 38 K (Furness et al., 2018).

If the substitution process persists, carriers experience excessive doping, leading to a decrease in Tc. La $^{3+}$ ions substitution with Sr $^{2+}$ ions results in the formation of a superconductor with high critical temperature. Regarding $Nd_{2-x}Ce_xCuO_4$, wherein Ce^{4+} ions substitute Nd^{3+} ions, the carriers in this substance are electrons, as opposed to holes (Krakauer & Pickett, 1988).

The structural composition of $Nd_{2-x}Ce_xCuO_4$ resembles that of $La_{2-x}Ba_xCuO_4$. However, the former structure is characterized by the absence of apical oxygen atoms positioned above and below the Cu atoms within the planar CuO_2 layers. The attainment of superconductivity with critical temperature (Tc) of 25 K is observed at a Ce 4+ ion concentration of x = 0.14. However, introducing further Ce^{4+} ions results in the disappearance of Tc. The coexistence of superconductivity and antiferromagnetism is observed in $La_{2-x}Ba_xCuO_4$ and $Nd_{2-x}Ce_xCuO_4$ (Feng et al., 2008).

The superconductor $YBa_2Cu_3O_{7-x}$, also called YBCO or Y_{123}, was the first to exhibit critical temperature (Tc) above that of liquid nitrogen (77K) and achieved a Tc of 93K. The present material exhibits a structural arrangement wherein the intercalation of Y atoms is observed between the two adjacent CuO_2 pyramid-shaped layers. The crystal exhibits a triple periodic arrangement wherein Y and Ba atoms are aligned in a Ba–Y–Ba configuration. Furthermore, one-dimensional chains of Cu–O–Cu are oriented along the b-axis of crystal (Feng et al., 2002). Chemical formula $YBa_2Cu_3O_{7-x}$ denotes significance of oxygen deficiency in facilitating process of hole implantation into CuO_2 layers. Degree of oxygen insufficiency is regulated by the influx and efflux of oxygen within CuO chains. The critical temperature (Tc) attains its peak value of 93 K in the vicinity of $x \approx 0$ in the Ortho I orthorhombic system. In comparison, it reduces to 60 K at $x \approx 0.5$ in the Ortho II orthorhombic system. Subsequently, the superconductivity ceases at $x \approx 0.7$ in the Tetra tetragonal system. The Ortho I crystal exhibits lattice constants of a = 0.382 nm, b = 0.388 nm, and c = 1.168 nm. Disparity between the lattice constants a and b induces a localized distortion, which is mitigated by the introduction of a twin (Oussena et al., 1996).

The Bi-based cuprate superconductors, namely $Bi_2Sr_2CaCu_2O_8$ (2212 phase), $Bi_2Sr_2CuO_6$ (2201 phase), and $(Bi, Pb)_2Sr_2Ca_2Cu_3O_{10}$ (2223 phase) are commonly utilized for wire applications. These materials exhibit critical temperatures of 30 K, 95 K, and110 K. Materials 2212 and 2223 involve addition of Sr and Ca, which are alkaline earth metals with bivalent ions, to 2201 simultaneously. This resulted in 2201 exhibiting a Tc exceeding 100 K, marking the first occurrence. The Bi-based superconductors exhibit distinct phases wherein perovskite layers are intercalated between Bi_2O_2 layers. Furthermore, number of CuO_2 layers in these phases increases sequentially. The crystal structures of superconductors based on Tl and Hg exhibit similarities (Krusin et al., 1994).

2.4.2.2. Electronic Structure

In order to comprehend the fundamental electronic configuration of high-temperature superconductor of the cuprate family, it is imperative to contemplate La_2CuO_4, which serves as the substrate crystal of $La_{2-x}Ba_xCuO_4$ that was initially discovered. The process of electric charge neutralization in a material involves the transfer of 2 electrons from the layer of La_2O_2 to CuO_2, considering the ions La^{3+} and O^{2-}. The resulting charge distribution of $(La_2O_2)^{2+}$ $(CuO_2)^{2-}$ is determined as average. In this context, let us examine the electronic configuration of the CuO_2 plane (Horide et al., 2008). The electronic configurations of Cu^{2+} characterize CuO_2 layer with 3d ^{9} and O^{2-} with 2p6. CuO_2 layer contains a Cu^{2+} ion with a hole in the d_{x2-y2} orbital. This hole facilitates the expansion of bonds towards four oxygen ions. Conversely, the electrons have filled all pσ orbitals of O_2^- ions that encircle the Cu $^{2+}$ ion. The d_{x2-y2} orbital exhibits a lower energy state than pσ orbital. If electron is forcibly confined within the d_{x2-y2} orbital, energy increases due to the Coulombic repulsion between electrons. The magnitude of the energy U is significant, measuring at 8 electron volts. The observed phenomenon can be attributed to electron correlation, as previously noted (Tanda et al., 1992).

Hence, despite the apparent lower energy of the d_{x2-y2} orbital compared to the p σ orbital, the electron occupancy of the former is incomplete, resulting in the Cu $^{2+}$ ion acquiring magnetic properties with a spin value of S = 1/2 (Reichhardt & Gronbech, 2001).

As an illustration, consider the partial substitution of La^{3+} ions with Ba^{2+} ions. Currently, the O^{2-} ions undergo electron removal, and resulting holes are introduced into CuO_2 layer. Electric conduction occurs due to unrestricted mobility of the injected holes within the crystal lattice. Additionally, the aperture nullifies the spin of the Cu^{2+} ion location through exchange interaction. The Cu^{2+} ions in CuO_2 layer exhibit antiferromagnetic interaction J, which is induced by the mutual influence of adjacent spins (Demirdis et al., 2013). The incorporation of a pair of apertures within the CuO_2 stratum results in the disturbance of the four antiferromagnetic linkages encircling a hole, owing to the annulment of spin instigated by the exchange interaction. This increase total energy of 8 J. Conversely, energy augmentation amounts to 7 J due to creation of two neighboring apertures. (Yuan et al., 2016).

Energy difference of 8 J–7 J can be postulated as cohesive force that binds the two orifices together. The Cooper pair associated with the hole is posited to possess distinct mechanisms from those of the BCS superconductors, albeit still considered a Cooper pair. The phenomenon of phase separation into distinct regions of hole and spin occurs under conditions of the low kinetic energy of hole. The phenomenon of superconductivity is believed to arise from a delicate equilibrium between attractive and kinetic energy of hole, coupled with the disappearance of antiferromagnetism due to thermal fluctuation. The text provides a qualitative account of emergence of superconductivity in cuprate superconductors, proposing superconductivity occurs near insulator (Xing et al., 2019).

2.4.2.3. Nisotropy and Vortex Lattice Melting

Cuprate superconductors exhibit a reflection of the CuO_2 layer structure, with a notable manifestation of two-dimensional traits in their diverse physical properties. The CuO_2 layer exhibits a significant difference between the out-of-plane resistivity (ρ_c) and in-plane resistivity (ρ_{ab}), with a ratio of ρc/ρab greatly exceeding 1. The implication is that the carriers exhibit restricted mobility along the c-a axis and possess a comparatively greater effective mass (mc) than their in-plane effective mass (mab). It is worth noting that cuprate superconductors belong to the category of type II superconductors (Xu et al., 2003). The subsequent equations illustrate the correlation between the effective mass ratio and the anisotropic parameter γa, coherence length ξ, upper critical field H_{c2}, and penetration depth λ.

$$(m_c/m_{ab})^{1/2} = \gamma_a = \xi_{ab}/\xi_c = H_{c2//}/H_{c2\perp} = \lambda_c/\lambda_{ab}, \qquad (42)$$

The symbol "//" denotes the circumstance in which a magnetic field is applied parallel to the layer. In contrast, the symbol "⊥" signifies the scenario in which the magnetic field is applied in a direction perpendicular to the layer. The derivation of the H_{c2} (θ) formula involves an examination of the angle θ that exists between the CuO_2 layer's parallel direction and the direction of the magnetic field (Kim et al., 2022).

$$H_{c2}(\theta) = H_{c2//} / (\cos^2\theta + \gamma_a^2 \sin^2\theta)^{1/2}. \tag{43}$$

The model demonstrating anisotropy is widely known as the effective mass model, as cited in reference The reported values of γ a for $YBa_2Cu_3O_{7-x}$ and $Bi_2Sr_2CaCu_2O_8$ are approximately 5 and 24, respectively. The material $YBa_2Cu_3O_{7-x}$ displays a disparity in coherence length, with ξc being shorter along c-axis lattice constant and ξab being longer than lattice constant of the a/b axes. The material $Bi_2Sr_2CaCu_2O_8$8 displays a notably brief coherence length, symbolized by ξc. Consequently, the confinement of the wave function of Cooper pairs within CuO $_2$ results in the manifestation of significant two-dimensional characteristics. The present study elucidates the anisotropic characteristics using the Lawrence-Doniach model (Pautrat et al., 2004;Sun et al., 2009).

Distinct features are observed in the high-temperature cuprate superconductor upon the imposition of a magnetic field. To provide an example:

- The pinning strength is relatively weak due to the small size of the cores of quantized vortices, which is a result of their limited coherence length;
- The ability of vortices to bend upon vertical penetration of the cuo the low superconducting coupling between these layers can explain 2 layers (Chen et al., 2009).
- As per the findings of a research paper, the Abrikosov lattice exhibits a tendency to undergo melting in the proximity of its critical temperature, owing to the impact of thermal energy. The cuprate superconductors demonstrate a noteworthy influence of thermal fluctuation, resulting in an ambiguous phase transition between normal conducting state and superconducting state.

Near critical field strength H_{c2}, amplitude of superconducting order parameter undergoes a process of evolution. At the same time, the phase experiences temporal variations, resulting in the emergence of a vortex liquid state. The vortex lattice in this state exhibits a lack of long-range order (Matsumoto et al., 2005).

However, a vortex lattice can exist in a solid state, commonly known as the vortex solid, where the phase is immobilized. This phenomenon takes place within a magnetic field range that is notably lower than upper critical field (H_{c2}). Intermediate region is where vortex solid transitions to vortex liquid state of vortex lattice occurs. The transition is a 1st-order transition. The melting lines for $Bi_2Sr_2CaCu_2O_8$ and $YBa_2Cu_3O_{7-x}$. The two-dimensional characteristics of liquid vortex state in $Bi_2Sr_2CaCu_2O_8$ result in notably wider region of manifestation. On the other hand, the vortex liquid phase exhibits a comparatively restricted region in $YBa_2Cu_3O_{7-x}$ material (Matsumoto & Mele, 2009).

2.4.2.4. Bose Glass and Vortex Glass

Achievement of the solid vortex state is contingent upon the confinement of discrete quantized vortices using crystalline defects in the superconductive substance. The solid vortex that arises from random distribution of pinning centers, often observed in presence of oxygen vacancies, is commonly called a vortex glass phase. Vortex glass exhibits a distinctive type of ordered state contingent upon the sample's condition, despite its absence of periodic long-range order akin to that of spin glasses. Immobilization of vortices characterizes this ordered state (Velez et al., 2008). Under such circumstances, the resistivity can be ascertained.

$$\rho = \lim_{J \to 0}(E/J) = 0 \qquad (44)$$

The symbol E is conventionally used to represent electric field, whereas current density is typically denoted by the symbol J. The state of superconductivity is achieved when the resistivity (ρ) reaches zero, which happens when the current density (J) approaches zero. Phase transition from the state of vortex liquid to that of vortex glass is categorized as 2ND-order transition. Furthermore, in scenarios where the crystal defect exhibits c-axis correlation, including columnar defects which emerge from heavy ion irradiation or twin boundaries, vortices experience pinning and stabilization by these defects upon application of a magnetic field parallel to c-axis. The Bose glass phase is distinguished by the stabilization of vortices via c-axis correlated defects. The effective handling of defects in high-Tc cuprate superconductors is paramount for their practical realization (Tung et al., 2006).

2.4.3. Other High-T_c Oxide Superconductors

To date, cuprate superconductors remain the only high-temperature superconducting materials identified, as previously discussed. The perplexing enigma lies in the exclusive manifestation of high-Tc solely in the cuprate superconductors. To tackle this matter, a deliberate endeavor has been made to investigate oxide superconductors that differ from the cuprates thoroughly. The preliminary inquiries conducted on $BaPb_{1-x}Bi_xO_3$ and $Ba_{1-x}K_xBiO_3$ can be considered as pioneering endeavors. Following this, Sr_2RuO, $Na_xCoO_2 \cdot yH_2O$, among other similar compounds, have been investigated for their remarkable superconducting characteristics that do not rely on the CuO $_2$ layer. Subsequent to unearthing La–O–Fe–P superconductors, a sequence of $LaO_{1-x}F_x$ and FeAs oxypnictides have been detected via persistent investigation for novel superconductors. The oxypnictides in question are composed of alternating layers of $LaO_{1-x}F_x$ and FeAs and exhibit a Tc of 26 K as of 2008. A study shows that the Tc value can be elevated to 56K by replacing La with Sr and Sm (Horide et al., 2013).

Crystal structure of $LaO_{1-x}F_x$ FeAs (1111 phase), a common iron-based superconductor. The compound above comprises multiple FeAs layers, wherein the FeAs layer is doped with electrons from the LaO layer. Furthermore, superconductivity is exhibited at a temperature of 38 K in ternary iron arsenide BaFe 2 As 2 (122 phase) through hole doping and potassium partially replacing barium [56]. The materials under consideration exhibit a comparable association with the superconductors $Nd_{2-x}Ce_xCuO_4$ and $La_{2-x}Ba_xCuO_4$.

Crystal structure of BaFe $_2$ As $_2$. In LnOFeAs (Ln: lanthanoid) system, the unit cell comprises a solitary layer of FeAs, whereas, in the B $BaFe_2As_2$ system, unit cell consists of two FeAs. Subsequently, FeSe (11 phases), which is an iron-based superconductor possessing a comparable structure and critical temperature (Tc) of 14 K, was discovered. Compounds in question exhibit anisotropic superconducting properties that are analogous to those observed in the extensively researched superconducting cuprate oxides. Anticipated are discoveries of additional high-Tc materials owing to the multiple combinations between carrier supply layer and superconducting layer in these substances (Muralidhar et al., 2002).

Despite not being oxides, $BaFe_2As_2$ and FeSe are speculated to possess a comparable superconductivity mechanism owing to their iron-based composition.

Similar to cuprate superconductors, the grain boundary of these materials may exhibit a limited superconducting current owing to their notably high μ_0H_{c2} and abbreviated coherence length. The comprehension acquired via examining cuprate superconductors will prove advantageous in the quest for a resolution to the problem (Hatano and Nelson, 1997).

2.5. CRITICAL CURRENTS AND VORTEX PINNING

Superconducting wires are capable of transmitting large amounts of current without experiencing any energy loss due to their electrical resistance being zero. Here, connection between vortex (flux pinning) and critical currents is briefly discussed.

2.5.1. Critical Current-Voltage and Currents Characteristics

2.5.1.1. Current–Voltage Characteristics

The capacity to conduct large currents through superconductors without encountering electrical resistance is a vital requirement for numerous applications. The critical current, denoted as I_c, represents the maximum attainable current. Furthermore, the density of critical current. The term " J_c " is commonly employed in the academic literature to denote superconductor's critical current density, defined as maximum current which can flow through unit cross-sectional area of material without any resistance. The current denoted by I_c is generated in response to the established voltage. Electrical resistivity standard of 10^{-14} Ω m and electric field standard of 1 μ V cm^{-1} are commonly utilized in determining I_c (Coffey & Clem, 1991). The phenomenon known as vortex creep, also referred to as flux creep, attains significant magnitude in the vicinity of I_c. Upon increasing the current, a distinct region emerges where Ohm's law is observed, while a vortex flow region manifests on side with elevated current. Power law model characterizes low voltage region beneath vortex flow area.

$$E(I) \propto (I/I_c)^n \quad (45)$$

The index is represented by symbol "n" in context of superconducting materials is utilized to characterize their properties. Material possessing a high n value is deemed homogeneous and of utmost importance for various applications (Llordes et al., 2012).

2.5.2. Depairing Current Density

Superconductor critical current density is then predicted. Disintegration of the Cooper pair occurs concomitantly with the attainment of the apex value, which is achieved upon the kinetic energy of pair reaching energy gap of 2Δ. This statement is incomplete and requires further context to be rewritten academically. Please provide more information or a complete,

$$(1/2)m\, v^{*}{}_{F}^{2} = 2\Delta \quad (46)$$

The expression "m* = 2 m" denotes the effective mass twice the rest mass, while "v F" represents the Fermi velocity. The equation J = e* n_s v_F is modified by substituting v F to obtain an approximate calculation. The transformation of Equation 31 in the GL theory yields a more accurate outcome as follows.

$$J_d = \Phi_0/(3\sqrt{3}\pi\mu_0\lambda^2\xi) \quad (47)$$

The abbreviation utilized for the current density at which the onset of depairing occurs is denoted as J $_d$. Under ideal conditions, maximum current density that superconductors can carry is generally within range of 10^5–10^6A cm^2. In contrast, copper wires have a maximum current density of approximately 10^2–10^3 A cm^{-2} under ideal conditions. Nevertheless, the attainment of theoretical limit of critical current density (J_c) in practical superconducting wires remains elusive. The phenomenon occurs because magnitude of the vortex pinning directly influences the production of voltage in the current-voltage curve (Obradors et al., 2012).

2.5.3. Vortex Pinning

2.5.3.1. Elementary Pinning Force

Phenomenon of vortex flow is observed in mixed state of type 2 superconductor, which is devoid of any crystal defects, when it is exposed to an electric current due to the action of the Lorentz force. Nevertheless, certain crystal imperfections that exist in superconducting materials generate pinning forces that hinder flow of vortices in opposition to Lorentz forces which are exerted on them. Vortex pinning is caused by imperfections known as pinning centers (Epstein & Baym, 1988). Pinning centers are classified into distinct categories based on their dimensions.

- Point defects, voids, vacancies in the oxygen chain, tiny precipitates, etc.;
- Columnar defect, dislocation, nanorod, etc.;
- The stacking fault, interface, planar precipitate, grain boundary, etc.

Multiple techniques facilitate the interaction between the flaw and the vortex. One may consider a superconductive material containing a solitary vortex distinct from a minor, customary conductive precipitation. The cylindrical core of the vortex experiences a loss of superconductivity, resulting in a zero-order parameter ψ at that location. Consequently, in cases where the vortex persists within the superconducting domain, the overall energy experiences an augmentation equivalent to condensation energy of core volume. Cancellation of energy penalty occurs when the vortex is positioned within the sediment, which is directly proportional to the condensation energy of the intersection volume between the normal core and precipitate (Osamura et al., 1986). Difference in energy levels leads to the emergence of the pinning energy U_p.

$$U_p = (1/2)\mu_0\pi\xi_0^2H_c^2d \quad (48)$$

The variable "d" represents the distance separating the vortex and the sediment. The dominant force responsible for pinning the precipitate, denoted as f_p, arises from the energy variation at its maximum point during the movement of the vortex across it. The following expression can approximate this force:

$$f_p \approx U_p\xi = (1\ 2)\mu\ \pi\xi_0H_c^2d \quad (49)$$

Apart from the customary precipitation, the grain boundary is a noteworthy pinning site. Within vicinity of grain boundary, electron undergoes scattering, leading to a decrease in its average distance traveled between collisions, commonly referred to as the mean free path. Coherence length, represented by symbol ξ, displays an inverse relationship with the length scale l, whereby a decrease in l leads to a proportional decrease in ξ. Upon the approach of the vortex to grain boundary, dimensions of normal core undergo a reduction, leading to a narrowing of the area of superconductivity loss. Sources document that the existence of a vortex within the grain boundary leads to an elevation in energy, that function as pining energy (Sneppen, 1992).

Vortex's typical core gives rise to fundamental interactions, which are concomitant with a magnetic interaction resulting from modification in dispersion of superconducting current encompassing the vortex. If a superconductor contains a planar precipitate with a thickness smaller than ξ, the superconducting current pattern surrounding the vortex will change as vortex approaches precipitate. This will eventually lead to Josephson vortex creation. Josephson vortex is distinguished by lack of a core, wherein solely the shielding current is present and extends along the planar precipitate, culminating in the magnetic flux's quantization. A planar precipitate with a slender structure has the potential to produce a pinning interaction which is nearly comparable to the core interaction. Resultant pinning force at the elementary level is widely believed to exhibit a high degree of resilience (Matsushita et al., 1986).

2.5.3.2. Vortex Lattice Elasticity

The emergence of triangular configuration of quantized vortices is attributed to repulsive interplay among these vortices in existence of a magnetic field, provided that intensity of field fulfills the criterion $H_{c1}<H<H_{c2}$ in type 2 superconductors. Elasticity of vortex lattice can be analyzed by considering it as a continuous body that is anisotropic in nature, as well as a crystal that is conventional in its characteristics. Given the alignment of the z-axis with the vortex direction, it follows that x-y and x-z planes will be orthogonal and parallel to vortex direction. The present analysis draws upon prior research and visual representation to demonstrate that the shear moduli C_{11}, C_{44}, and C_{66} correspond to discrete mechanical deformations, specifically uniaxial compression in the x-y plane, bending in the x-z plane, and transverse shear in the x-y plane, respectively. The determination of elastic moduli of C_{11}, C_{44}, and C_{66} can be achieved by utilizing the principles of electromagnetism and thermodynamics. Expressions for these moduli can be derived accordingly (Mele et al., 2008).

$$C_{11}\,C_{44} = B^2\,\mu_0 \text{ and,} \tag{50}$$

$$C_{66}(B\Phi_0\,16\pi\lambda^2\mu_0)(1-1\ 2K^2)(1-h^2)(1-0\ 58.h+0\ 29.h^2) \tag{51}$$

The variable h may be defined as the quotient of H divided by H_{c2}, denoted as $h = H/H_{c2}$. As reported in reference, it is generally observed that the value of C_{66} is significantly smaller than that of C_{11} and C_{44}. The behavior of the elastic constants C_{11}, C_{44}, and C_{66} is dependent on the magnetic field. Specifically, C_{11} and C_{44} display an increasing trend as the magnetic field intensifies, whereas C_{66} initially increases but subsequently decreases as magnetic field approaches H_{c2}. Vortex lattice displays notable resistance to compression and bending when subjected to intense magnetic fields while demonstrating reduced resistance to shearing. The assessment of vortex pinning necessitates the careful consideration of the elasticity of the vortex lattice (Ochiai & Osamura, 1986).

2.5.4. Global Pinning Force

The state of balance between Lorentz force which operates on individual vortex and pinning force that pinning center applies leads to immobilization of the vortex until the point where Lorentz force cross critical pinning force (f_p). The concept of "global pinning force" (f_p) pertains to highest possible pinning force per unit volume achievable by employing multiple pinning centers.

$\mathbf{F}_p = \mathbf{J}_c \times \mathbf{B}$ (52)

The mathematical representation of $\mathbf{F}_p$ can be expressed as a composite function that includes density of pins N_p and fundamental elementary pinning force f_p. Determination of f_p under particular magnetic field and temperature circumstances is frequently referred to as a summation conundrum, which is dependent on specified values of f_p and N_p. The described situation can be categorized as many-body problem, where elastic energy present among vortices opposes pinning energy. Moreover, incorporating thermal energy in high-temperature T_c superconductors intensifies problem, making it more complex (Juang et al., 1992).

First, suppose a situation where pinning mechanism remains undisturbed by the vortex lattice. This can be attributed to the low value of f_p and high rigidity of vortex lattice. The vortex lattice encounters minimal stochastic forces if the pinning centers are distributed stochastically. Therefore, the force generated is commonly known as the resultant force.

$$F_p = |\mathbf{F}_p| = \sum f_p = 0 \qquad (53)$$

Precisely, the magnitude of J_c is equivalent to null. The term "weak pinning limit" is a widely used in scholarly works to describe the phenomenon above. (Baumgartner et al., 2013). Consequently, a correlation is established between the value of f_p.

$$F_p = |\mathbf{F}_p| = \sum f_p = N_p f_p \qquad (54)$$

The current methodology employs the direct summation method, whereby the upper threshold of J_c achievable via pinning is delimited by the interval. The critical domain of real superconductors lies between the aforementioned limits. Additionally, if the magnitude of F_p is small enough for vortex lattice to display collective behavior, it is possible to estimate numerical F_p value using pinning model. The abovementioned model is commonly employed to explain vortex pinning phenomena observed in high-temperature superconductors (Chang et al., 1992).

2.5.4.1. Scaling Rule

The scaling rule can demonstrate the correlation between Fp and variations in magnetic field and temperature.

$$F_p = KH h_{c\,2}^{m\,p}(1-h)^q \qquad (55)$$

The equation in question incorporates the constants K, m, p, and q, which have been ascertained through empirical methods. Furthermore, the variable denoted as h is established as the quotient of H divided by H_{c2}, expressed as, $h = H/H_{c2}$. The range of 1.5 to 2.5 is frequently observed for the value of m in metallic superconductors. Moreover, it is widely acknowledged in the scientific community that the typical values of the parameters and "q" in practical Nb-Ti superconducting wires are approximately 0.5–1 and 1, respectively. In contrast, conventional superconducting wires including Nb_3Sn demonstrate values of p approximately between 0.5 and 2 (Miura et al., 1992).

The magnetic field dependence of F_p. Relationship denoted by $F_p \propto (1 - h)$ is known as the saturation characteristics and has been observed in various materials, including Nb_3Sn. The decrease in the elastic coefficient C_{66} is observed in response to magnetic field dependence of $(1 - h)^2$, even when vortex lattice is pinned.

As a result, the Lorentz force plays a crucial role in inducing shear deformation in vortex lattice, which restricts the flow of vortices produced and critical current density (J_c). In literature, correlation between $F_p \propto (1 - h)$ is commonly regarded as non-saturated attribute. The characteristics above holds substantial importance within superconducting coils, as it indicates that the critical current density (J_c) sustains a notable level despite exposure to high magnetic fields. This phenomenon becomes evident when plasma frequency value (f_p) is comparatively elevated (Matsushita, 1981).

The irreversibility line Hirr is utilized instead of H_{c2} to investigate the scaling rule in cuprate superconductors, as it serves as the delineation between $J_c = 0$ and $Jc \neq 0$. This methodology has been recorded in scholarly works. In the region situated beneath the H_{irr} curve, which is typified by low temperature and low filling, flux pinning is evident (Weygand et al., 1999). It leads to a critical current density (Jc) that isn't equal to 0. Metallic superconductors exhibit a H_{irr} curve positioned below its H_{c2}, whereas cuprate superconductors demonstrate a Hirr notably lower than its H_{c2}. The quantized vortices in cuprate superconductors display effortless mobility due to thermal fluctuations. As a result, when in close proximity to H_{c2}, energy dissipation takes place with ease, ultimately reducing Jc to zero. However, empirical evidence has demonstrated that F p behavior across a broad range of magnetic and temperature fields can also be explained by scaling principle that utilizes Hirr instead of H_{c2} (Balankin et al., 2003).

Activity 2.1

You are attending a physics conference where experts in the field of high-temperature superconductivity are presenting their research. As a participant, your task is to engage in a debate session discussing the various theories proposed to explain the mechanism behind high-temperature superconductivity.

2.6. SUMMARY

Chapter 2 provides a comprehensive overview of the general theory surrounding high-TC (critical temperature) superconductors. It begins with an introduction to the chapter's topics and then delves into the fundamental properties of superconductors, including the superconducting phenomenon, the connection between superconductivity and quantum mechanics, and the thermodynamics of superconductivity. The chapter also explores the behavior of superconductors in magnetic fields.

The discussion then moves on to different types of superconducting materials, including superconducting metals and alloys, superconducting compounds, and organic superconductors. The chapter specifically focuses on high-TC oxide superconductors, providing insights into early oxide superconductors, cuprate superconductors with their unique layered crystal structures, and other high-TC oxide superconductors.

Additionally, the chapter covers critical currents and vortex pinning, explaining concepts such as critical current-voltage characteristics, depairing current density, vortex pinning, and global pinning force.

REVIEW QUESTIONS

1. Why is understanding the quantum mechanical aspect of superconductivity important in studying high-TC superconductors?
2. What are the fundamental properties of superconductors discussed in this chapter?
3. Explain the relationship between superconductivity and thermodynamics.
4. How do superconductors behave in the presence of a magnetic field?

5. What are some examples of superconducting materials mentioned in this chapter?
6. Describe the characteristics and properties of high-TC oxide superconductors.
7. What are cuprate superconductors, and why are their layered crystal structures significant?
8. What is the critical current and how does it relate to superconductors?
9. Explain the concept of vortex pinning and its importance in high-TC superconductors.
10. How is the global pinning force relevant to the behavior of superconducting vortices?

REFERENCES

1. Abrikosov, A. A. (1957). The magnetic properties of superconducting alloys. Journal of Physics and Chemistry of Solids, 2(3), 199–208.
2. Abrikosov, A. A., & Gor'Kov, L. P. (1958). On the theory of superconducting alloys. I. The electrodynamics of alloys at absolute zero. Zhur. Eksptl.' i Teoret. Fiz., 35.
3. Abrikosov, A. A., & Gor'kov, L. P. (1960). Contribution to the theory of superconducting alloys with paramagnetic impurities. Zhur. Eksptl.' i Teoret. Fiz., 39.
4. Afshar, S. S., Flores, E., McDonald, K. F., & Knoesel, E. (2007). Paradox in wave-particle duality. Foundations of Physics, 37, 295–305.
5. Albash, T., & Johnson, C. V. (2008). A holographic superconductor in an external magnetic field. Journal of High Energy Physics, 2008(09), 121.
6. Ambegaokar, V., & Halperin, B. I. (1969). Voltage due to thermal noise in the dc Josephson effect. Physical Review Letters, 22(25), 1364.
7. Ando, Y., Kurita, Y., Komiya, S., Ono, S., & Segawa, K. (2004). Evolution of the Hall coefficient and the peculiar electronic structure of the cuprate superconductors. Physical Review Letters, 92(19), 197001.
8. Angelo, R. M., & Ribeiro, A. D. (2015). Wave–particle duality: An information-based approach. Foundations of Physics, 45, 1407–1420.
9. Arndt, M., Nairz, O., Vos-Andreae, J., Keller, C., Van der Zouw, G., & Zeilinger, A. (1999). Wave–particle duality of C_{60} molecules. Nature, 401(6754), 680–682.
10. Aspect, A., & Grangier, P. (1987). Wave-particle duality for single photons. Hyperfine Interactions, 37(1–4), 1–17.
11. Aston, D. R., Dubeck, L. W., & Rothwarf, F. (1971). "Intermediate Mixed" State of Type-II Superconductors. Physical Review B, 3(7), 2231.
12. Babaev, E., & Speight, M. (2005). Semi-Meissner state and neither type-I nor type-II superconductivity in multicomponent superconductors. Physical Review B, 72(18), 180502.
13. Balachandran, A. P., & Teotonio-Sobrinho, P. (1993). The Edge states of the BF system and the London equations. International Journal of Modern Physics A, 8(04), 723–752.
14. Balankin, A. S., Susarrey, O., & Gonzáles, J. M. (2003). Scaling properties of pinned interfaces in fractal media. Physical Review Letters, 90(9), 096101.
15. Baumgartner, T., Eisterer, M., Weber, H. W., Flükiger, R., Scheuerlein, C., & Bottura, L. (2013). Effects of neutron irradiation on pinning force

scaling in state-of-the-art Nb_3Sn wires. Superconductor Science and Technology, 27(1), 015005.

16. Bergeret, F. S., Volkov, A. F., & Efetov, K. B. (2005). Odd triplet superconductivity and related phenomena in superconductor-ferromagnet structures. Reviews of Modern Physics, 77(4), 1321.
17. Bloch, F. (1968). Simple interpretation of the Josephson effect. Physical Review Letters, 21(17), 1241.
18. Böhm, T., Kretzschmar, F., Baum, A., Rehm, M., Jost, D., Hosseinian Ahangharnejhad, R., & Hackl, R. (2018). Microscopic origin of Cooper pairing in the iron-based superconductor $Ba_{1-x}K_xFe_2As_2$. NPJ Quantum Materials, 3(1), 48.
19. Brandt, E. H. (1995). The flux-line lattice in superconductors. Reports on Progress in Physics, 58(11), 1465.
20. Brandt, E. H., & Indenbom, M. (1993). Type-II-superconductor strip with current in a perpendicular magnetic field. Physical Review B, 48(17), 12893.
21. Buzea, C., & Robbie, K. (2004). Assembling the puzzle of superconducting elements: a review. Superconductor Science and Technology, 18(1), R1.
22. Chang, H. L., Wang, C., Juang, J. Y., Uen, T. M., & Gou, Y. S. (1992). The Scaling Behaviors of Pinning in TlBaCaCuO Superconducting Thin Films. Chinese Journal of Physics, 30(4), 467–475.
23. Chen, X., Xia, Y., Huang, Q., Li, Z., Mei, A., Hu, Y., & Han, H. (2021). Tailoring the dimensionality of hybrid perovskites in mesoporous carbon electrodes for type-II band alignment and enhanced performance of printable hole-conductor-free perovskite solar cells. Advanced Energy Materials, 11(18), 2100292.
24. Chen, Z., Kametani, F., Chen, Y., Xie, Y., Selvamanickam, V., & Larbalestier, D. C. (2009). A high critical current density MOCVD coated conductor with strong vortex pinning centers suitable for very high field use. Superconductor Science and Technology, 22(5), 055013.
25. Coffey, M. W., & Clem, J. R. (1991). Unified theory of effects of vortex pinning and flux creep upon the rf surface impedance of type-II superconductors. Physical Review Letters, 67(3), 386.
26. Cyrot, M. (1973). Ginzburg-Landau theory for superconductors. Reports on Progress in Physics, 36(2), 103.
27. Damascelli, A., Hussain, Z., & Shen, Z. X. (2003). Angle-resolved photoemission studies of the cuprate superconductors. Reviews of Modern Physics, 75(2), 473.

28. Demirdiş, S., Fasano, Y., Kasahara, S., Terashima, T., Shibauchi, T., Matsuda, Y., & Van Der Beek, C. J. (2013). Disorder, critical currents, and vortex pinning energies in isovalently substituted BaFe 2 (As 1-x P x) 2. Physical Review B, 87(9), 094506.
29. Dimitrova, T. L., & Weis, A. (2008). The wave-particle duality of light: A demonstration experiment. American Journal of Physics, 76(2), 137–142.
30. Duan, Y., Zhang, H., & Li, S. (1998). Topological structure of the London equation. Physical Review B, 58(1), 125.
31. Edwards, W. F. (1981). Classical derivation of the London equations. Physical Review Letters, 47(26), 1863.
32. Ekin, J. W. (1984). Strain effects in superconducting compounds. Advances in Cryogenic Engineering Materials: Volume 30, 823–836.
33. Emery, V. J., & Kivelson, S. A. (1993). Frustrated electronic phase separation and high-temperature superconductors. Physica C: Superconductivity, 209(4), 597–621.
34. Epstein, R. I., & Baym, G. (1988). Vortex pinning in neutron stars. The Astrophysical Journal, 328, 680–690.
35. Feng, D. L., Damascelli, A., Shen, K. M., Motoyama, N., Lu, D. H., Eisaki, H., & Shen, Z. X. (2002). Electronic Structure of the Trilayer Cuprate Superconductor Bi 2 Sr 2 Ca 2 Cu 3 O 1 0+ δ. Physical Review Letters, 88(10), 107001.
36. Feng, S., Guo, H., Lan, Y., & Cheng, L. (2008). Electronic structure of kinetic energy driven cuprate superconductors. International Journal of Modern Physics B, 22(22), 3757–3811.
37. Fradkin, E., Kivelson, S. A., & Tranquada, J. M. (2015). Colloquium: Theory of intertwined orders in high temperature superconductors. Reviews of Modern Physics, 87(2), 457.
38. Franke, K. J., Schulze, G., & Pascual, J. I. (2011). Competition of superconducting phenomena and Kondo screening at the nanoscale. Science, 332(6032), 940–944.
39. Furness, J. W., Zhang, Y., Lane, C., Buda, I. G., Barbiellini, B., Markiewicz, R. S., & Sun, J. (2018). An accurate first-principles treatment of doping-dependent electronic structure of high-temperature cuprate superconductors. Communications Physics, 1(1), 11.
40. Gabovich, A. M., Voitenko, A. I., & Ausloos, M. (2002). Charge-and spin-density waves in existing superconductors: competition between Cooper pairing and Peierls or excitonic instabilities. Physics Reports, 367(6), 583–709.

41. Gantmakher, V. F., & Dolgopolov, V. T. (2010). Superconductor–insulator quantum phase transition. Physics-Uspekhi, 53(1), 1.
42. Gatti, A., Brambilla, E., & Lugiato, L. A. (2003). Entangled imaging and wave-particle duality: from the microscopic to the macroscopic realm. Physical Review Letters, 90(13), 133603.
43. Geerligs, L. J., Anderegg, V. F., Romijn, J., & Mooij, J. E. (1990). Single Cooper-pair tunneling in small-capacitance junctions. Physical Review Letters, 65(3), 377.
44. Goodman, B. B. (1964). Type II or London superconductors. Reviews of Modern Physics, 36(1), 12.
45. Gotoda, H., Osamura, K., Furusaka, M., Arai, M., Suzuki, J., Lee, P. J., & Monju, Y. (1989). Influence of microstructure on the flux pinning in Nb-Ti multifilamentary superconducting wires. Philosophical Magazine B, 60(6), 819–835.
46. Hamilton, D. C., Raub, C. J., Matthias, B. T., Corenzwit, E., & Hull Jr, G. W. (1965). Some new superconducting compounds. Journal of Physics and Chemistry of Solids, 26(3), 665–667.
47. Hatano, N., & Nelson, D. R. (1997). Vortex pinning and non-Hermitian quantum mechanics. Physical Review B, 56(14), 8651.
48. Hendry, J. (1980). The development of attitudes to the wave-particle duality of light and quantum theory, 1900–1920. Annals of Science, 37(1), 59–79.
49. Herbut, I. F. (2005). Effective theory of high-temperature superconductors. Physical Review Letters, 94(23), 237001.
50. Hirsch, J. E., Maple, M. B., & Marsiglio, F. (2015). Superconducting materials classes: Introduction and overview. Physica C: Superconductivity and Its Applications, 514, 1–8.
51. Hor, P. H., Gao, L., Meng, R. L., Huang, Z. J., Wang, Y. Q., Forster, K., & Torng, C. J. (1987). High-pressure study of the new Y-Ba-Cu-O superconducting compound system. Physical Review Letters, 58(9), 911.
52. Horide, T., Kawamura, T., Matsumoto, K., Ichinose, A., Yoshizumi, M., Izumi, T., & Shiohara, Y. (2013). Jc improvement by double artificial pinning centers of $BaSnO_3$ nanorods and Y_2O_3 nanoparticles in $YBa_2Cu_3O_7$ coated conductors. Superconductor Science and Technology, 26(7), 075019.
53. Horide, T., Matsumoto, K., Mele, P., Ichinose, A., Kita, R., Mukaida, M., & Horii, S. (2008). The crossover from the vortex glass to the Bose glass in nanostructured $YBa_2Cu_3O_{7-x}$ films. Applied Physics Letters, 92(18), 182511.

54. Hott, R., Kleiner, R., Wolf, T., & Zwicknagl, G. (2013). Review on superconducting materials. arXiv Preprint arXiv:1306.0429.
55. Jeanneret, B., & Benz, S. P. (2009). Application of the Josephson effect in electrical metrology. The European Physical Journal Special Topics, 172(1), 181–206.
56. Juang, J. Y., Wang, S. J., Uen, T. M., Gou, Y. S., Chang, H. L., & Wang, C. (1992). Low-field scaling behaviors of global flux pinning in Tl-Ba-Ca-Cu-O thin films. Physical Review B, 46(2), 1188.
57. Kim, J. H., Hidayati, R., Jung, S. G., Salawu, Y. A., Kim, H. J., Yun, J. H., & Rhyee, J. S. (2022). Enhancement of critical current density and strong vortex pinning in high entropy alloy superconductor Ta1/6Nb2/6Hf1/6Zr1/6Ti1/6 synthesized by spark plasma sintering. Acta Materialia, 232, 117971.
58. Kivelson, S. A., & Fradkin, E. (2005). How optimal inhomogeneity produces high temperature superconductivity. arXiv preprint cond-mat/0507459.
59. Kolesov, R., Grotz, B., Balasubramanian, G., Stöhr, R. J., Nicolet, A. A., Hemmer, P. R., & Wrachtrup, J. (2009). Wave–particle duality of single surface plasmon polaritons. Nature Physics, 5(7), 470–474.
60. Krakauer, H., & Pickett, W. E. (1988). Effect of bismuth on high-T c cuprate superconductors: Electronic structure of Bi 2 Sr 2 CaCu 2 O 8. Physical Review Letters, 60(16), 1665.
61. Kruchinin, S. P. (2014). Physics of high-Tc superconductors. Reviews in Theoretical Science, 2(2), 124–145.
62. Krusin-Elbaum, L., Civale, L., Blatter, G., Marwick, A. D., Holtzberg, F., & Feild, C. (1994). Bose-glass melting in YBaCuO crystals with correlated disorder. Physical Review Letters, 72(12), 1914.
63. Lake, B., Rønnow, H. M., Christensen, N. B., Aeppli, G., Lefmann, K., McMorrow, D. F., & Mason, T. E. (2002). Antiferromagnetic order induced by an applied magnetic field in a high-temperature superconductor. Nature, 415(6869), 299–302.
64. Larbalestier, D., Gurevich, A., Feldmann, D. M., & Polyanskii, A. (2001). High-T c superconducting materials for electric power applications. Nature, 414(6861), 368–377.
65. Lee, P. A. (1993). Localized states in a d-wave superconductor. Physical Review Letters, 71(12), 1887.
66. Li, P., Abraimov, D., Xu, A., & Larbalestier, D. (2011). Observation of important current-limiting defects in a recent high pinning force MOCVD

IBAD-MgO coated conductor. Superconductor Science and Technology, 25(2), 025002.

67. Likharev, K. K. (1983). True quantum-mechanical macroscopic effects in weak superconductivity. Soviet Physics Uspekhi, 26(1), 87.

68. Linder, J., & Sudbø, A. (2007). Quantum transport in noncentrosymmetric superconductors and thermodynamics of ferromagnetic superconductors. Physical Review B, 76(5), 054511.

69. Ling, C. D., Schmid, S., Blanchard, P. E., Petricek, V., McIntyre, G. J., Sharma, N., & Withers, R. L. (2013). A (3+ 3)-dimensional "hypercubic" oxide-ionic conductor: type II Bi_2O_3–Nb_2O_5. Journal of the American Chemical Society, 135(17), 6477–6484.

70. Liu, M. (1998). Rotating superconductors and the frame-independent London equation. Physical Review Letters, 81(15), 3223.

71. Liu, Y., Haviland, D. B., Nease, B., & Goldman, A. M. (1993). Insulator-to-superconductor transition in ultrathin films. Physical Review B, 47(10), 5931.

72. Llordes, A., Palau, A., Gázquez, J., Coll, M., Vlad, R., Pomar, A., & Obradors, X. (2012). Nanoscale strain-induced pair suppression as a vortex-pinning mechanism in high-temperature superconductors. Nature Materials, 11(4), 329–336.

73. London, F., & London, H. (1935). The electromagnetic equations of the supraconductor. Proceedings of the Royal Society of London. Series A-Mathematical and Physical Sciences, 149(866), 71–88.

74. Lowell, J., & Sousa, J. B. (1970). Mixed-state thermal conductivity of type II superconductors. Journal of Low Temperature Physics, 3, 65–87.

75. Maeda, H., Tanaka, Y., Fukutomi, M., & Asano, T. (1988). A new high-Tc oxide superconductor without a rare earth element. Japanese Journal of Applied Physics, 27(2A), L209.

76. Maki, K. (1964). The magnetic properties of superconducting alloys I. Physics Physique Fizika, 1(1), 21.

77. Matsumoto, K., & Mele, P. (2009). Artificial pinning center technology to enhance vortex pinning in YBCO coated conductors. Superconductor Science and Technology, 23(1), 014001.

78. Matsumoto, K., Horide, T., Ichinose, A., Horii, S., Yoshida, Y., & Mukaida, M. (2005). Critical current control in $YBa_2Cu_3O_{7-\delta}$ films using artificial pinning centers. Japanese Journal of Applied Physics, 44(1L), L246.

79. Matsushita, T. (1981). Pinning force of a nonideal superconductor containing pins with large interaction forces. I. high pin density. Japanese

Journal of Applied Physics, 20(6), 1153.

80. Matsushita, T., & Yamafuji, K. (1981). Pinning force density in nonideal superconductors with weak pins. Journal of the Physical Society of Japan, 50(1), 38–46.
81. Matsushita, T., Itoh, M., Kikitsu, A., & Miyamoto, Y. (1986). Elastic and plastic behavior of a fluxoid lattice in the saturation region of the global pinning force in superconducting Nb-Ta. Physical Review B, 33(5), 3134.
82. Matthias, B. T., Compton, V. B., & Corenzwit, E. (1961). Some new superconducting compounds. Journal of Physics and Chemistry of Solids, 19(1–2), 130–133.
83. Matthias, B. T., Corenzwit, E., & Miller, C. E. (1954). Superconducting compounds. Physical Review, 93(6), 1415.
84. Mele, P., Matsumoto, K., Horide, T., Ichinose, A., Mukaida, M., Yoshida, Y., & Kita, R. (2008). Ultra-high flux pinning properties of $BaMO_3$-doped $YBa_2Cu_3O_{7-x}$ thin films (M= Zr, Sn). Superconductor Science and Technology, 21(3), 032002.
85. Mele, P., Matsumoto, K., Horide, T., Ichinose, A., Mukaida, M., Yoshida, Y., & Kita, R. (2007). Incorporation of double artificial pinning centers in $YBa_2Cu_3O_{7-d}$ films. Superconductor Science and Technology, 21(1), 015019.
86. Milošević, M. V., & Geurts, R. (2010). The Ginzburg–Landau theory in application. Physica C: Superconductivity, 470(19), 791–795.
87. Miura, O., Matsumoto, K., Tanaka, Y., Yamafuji, K., Harada, N., Iwakuma, M., & Matsushita, T. (1992). Pinning characteristics in multifilamentary Nb□ Ti superconducting wires with submicrometre filaments introduced artificial pinning centres. Cryogenics, 32(3), 315–322.
88. Murakami, M., Gotoh, S., Koshizuka, N., Tanaka, S., Matsushita, T., Kambe, S., & Kitazawa, K. (1990). Critical currents and flux creep in melt processed high Tc oxide superconductors. Cryogenics, 30(5), 390–396.
89. Muralidhar, M., Sakai, N., Chikumoto, N., Jirsa, M., Machi, T., Nishiyama, M., & Murakami, M. (2002). New type of vortex pinning structure effective at very high magnetic fields. Physical Review Letters, 89(23), 237001.
90. Musterd, M., van Steijn, V., Kleijn, C. R., & Kreutzer, M. T. (2014). Droplets on inclined plates: Local and global hysteresis of pinned capillary surfaces. Physical Review Letters, 113(6), 066104.
91. Nakamura, T., Itoh, Y., Yoshikawa, M., Oka, T., & Uzawa, J. (2007). Development of a superconducting magnet for nuclear magnetic resonance

using bulk high-temperature superconducting materials. Concepts in Magnetic Resonance Part B: Magnetic Resonance Engineering: An Educational Journal, 31(2), 65–70.

92. Norman, M. R., & Pépin, C. (2003). The electronic nature of high temperature cuprate superconductors. Reports on Progress in Physics, 66(10), 1547.

93. Norman, M. R., MacDonald, A. H., & Akera, H. (1995). Magnetic oscillations and quasiparticle band structure in the mixed state of type-II superconductors. Physical Review B, 51(9), 5927.

94. Obradors, X., Puig, T., Ricart, S., Coll, M., Gazquez, J., Palau, A., & Granados, X. (2012). Growth, nanostructure and vortex pinning in superconducting $YBa_2Cu_3O_7$ thin films based on trifluoroacetate solutions. Superconductor Science and Technology, 25(12), 123001.

95. Ochiai, S., & Osamura, K. (1986). Influence of grain size and upper critical magnetic field on global pinning force of bronze-processed Nb_3Sn compound. Acta Metallurgica, 34(12), 2425–2433.

96. Ogata, M., & Fukuyama, H. (2008). The t–J model for the oxide high-Tc superconductors. Reports on Progress in Physics, 71(3), 036501.

97. Osamura, K., Ochiai, S., Kondo, S., Namatame, M., & Nosaki, M. (1986). Influence of third elements on growth of Nb 3 Sn compounds and on global pinning force. Journal of Materials Science, 21, 1509–1516.

98. Osamura, K., Takayama, T., & Ochiai, S. (1989). Weak flux pinning at grain boundaries in Ag sheathed $Ba_2YCu_3O_{6+x}$ tapes. Applied Physics Letters, 55(4), 396–398.

99. Oussena, M., De Groot, P. A. J., Deligiannis, K., Volkozub, A. V., Gagnon, R., & Taillefer, L. (1996). Vortex pinning by competing disorder: Bose-glass to vortex-glass crossover. Physical Review Letters, 76(14), 2559.

100. Pal, B., Chakraborty, A., Sivakumar, P. K., Davydova, M., Gopi, A. K., Pandeya, A. K., & Parkin, S. S. (2022). Josephson diode effect from Cooper pair momentum in a topological semimetal. Nature Physics, 18(10), 1228–1233.

101. Park, C., & Snyder, R. L. (1995). Structures of High-Temperature Cuprate Superconductors. Journal of the American Ceramic Society, 78(12), 3171–3194.

102. Patra, S., & Lal, S. (2021). Origin of topological order in a Cooper-pair insulator. Physical Review B, 104(14), 144514.

103. Pautrat, A., Scola, J., Goupil, C., Simon, C., Villard, C., Domengès, B., & Méchin, L. (2004). Quantitative analysis of the critical current due

to vortex pinning by surface corrugation. Physical Review B, 69(22), 224504.

104. Pippard, A. B., & Bragg, W. L. (1953). An experimental and theoretical study of the relation between magnetic field and current in a superconductor. Proceedings of the Royal Society of London. Series A. Mathematical and Physical Sciences, 216(1127), 547–568.
105. Reichhardt, C., & Grønbech-Jensen, N. (2001). Critical currents and vortex states at fractional matching fields in superconductors with periodic pinning. Physical Review B, 63(5), 054510.
106. Rosenstein, B., & Li, D. (2010). Ginzburg-Landau theory of type II superconductors in magnetic field. Reviews of Modern Physics, 82(1), 109.
107. Ross, D. K. (1983). The London equations for superconductors in a gravitational field. Journal of Physics A: Mathematical and General, 16(6), 1331.
108. Ruan, W., Li, X., Hu, C., Hao, Z., Li, H., Cai, P., & Wang, Y. (2018). Visualization of the periodic modulation of Cooper pairing in a cuprate superconductor. Nature Physics, 14(12), 1178–1182.
109. Saito, G., & Yoshida, Y. (2011). Organic superconductors. The Chemical Record, 11(3), 124–145.
110. Scalapino, D. J., White, S. R., & Zhang, S. (1993). Insulator, metal, or superconductor: The criteria. Physical Review B, 47(13), 7995.
111. Scanlan, R. M., Malozemoff, A. P., & Larbalestier, D. C. (2004). Superconducting materials for large scale applications. Proceedings of the IEEE, 92(10), 1639–1654.
112. Schuller, I. K., & Jorgensen, J. D. (1989). Structure of high Tc oxide superconductors. MRS Bulletin, 14(1), 27–30.
113. Schwindt, P. D., Kwiat, P. G., & Englert, B. G. (1999). Quantitative wave-particle duality and nonerasing quantum erasure. Physical Review A, 60(6), 4285.
114. Shurakov, A., Lobanov, Y., & Goltsman, G. (2015). Superconducting hot-electron bolometer: from the discovery of hot-electron phenomena to practical applications. Superconductor Science and Technology, 29(2), 023001.
115. Skalski, S., Betbeder-Matibet, O., & Weiss, P. R. (1964). Properties of superconducting alloys containing paramagnetic impurities. Physical Review, 136(6A), A1500.
116. Sneppen, K. (1992). Self-organized pinning and interface growth in a

random medium. Physical Review Letters, 69(24), 3539.

117. Sun, Y., Pyon, S., Tamegai, T., Kobayashi, R., Watashige, T., Kasahara, S., & Shibauchi, T. (2015). Critical current density, vortex dynamics, and phase diagram of single-crystal FeSe. Physical Review B, 92(14), 144509.
118. Sweedler, A. R., Cox, D. E., & Moehlecke, S. (1978). Neutron irradiation of superconducting compounds. Journal of Nuclear Materials, 72(1–2), 50–69.
119. Tachiki, M., & Takahashi, S. (1989). Strong vortex pinning intrinsic in high-Tc oxide superconductors. Solid State Communications, 70(3), 291–295.
120. Tanda, S., Ohzeki, S., & Nakayama, T. (1992). Bose glass-vortex–glass phase transition and dynamics scaling for high-T_cNd_{2-x} Ce_xCuO_4 thin films. Physical Review Letters, 69(3), 530.
121. Tarantini, C., Lee, S., Kametani, F., Jiang, J., Weiss, J. D., Jaroszynski, J., & Larbalestier, D. C. (2012). Artificial and self-assembled vortex-pinning centers in superconducting Ba $(Fe_{1-x}Co_x)_2$ As_2 thin films as a route to obtaining very high critical-current densities. Physical Review B, 86(21), 214504.
122. Tarascon, J. M., McKinnon, W. R., Barboux, P., Hwang, D. M., Bagley, B. G., Greene, L. H., & Giroud, M. (1988). Preparation, structure, and properties of the superconducting compound series $Bi_2Sr_2Ca_{n-1}$ Cu n O y with n= 1, 2, and 3. Physical Review B, 38(13), 8885.
123. Tomita, M., & Murakami, M. (2003). High-temperature superconductor bulk magnets that can trap magnetic fields of over 17 tesla at 29 K. Nature, 421(6922), 517–520.
124. Tsuei, C. C., & Kirtley, J. R. (2000). Pairing symmetry in cuprate superconductors. Reviews of Modern Physics, 72(4), 969.
125. Tung, S., Schweikhard, V., & Cornell, E. A. (2006). Observation of vortex pinning in Bose-Einstein condensates. Physical Review Letters, 97(24), 240402.
126. Ubaldini, A., Giannini, E., Senatore, C., & Van Der Marel, D. (2010). BiOCuS: A new superconducting compound with oxypnictide-related structure. Physica C: Superconductivity and Its Applications, 470, S356–S357.
127. Urayama, H., Yamochi, H., Saito, G., Nozawa, K., Sugano, T., Kinoshita, M., & Tanaka, J. (1988). A new ambient pressure organic superconductor based on BEDT-TTF with Tc higher than 10 K (Tc= 10.4 K). Chemistry Letters, 17(1), 55–58.

128. Usadel, K. D. (1970). Generalized diffusion equation for superconducting alloys. Physical Review Letters, 25(8), 507.

129. Valles Jr, J. M., Dynes, R. C., & Garno, J. P. (1992). Electron tunneling determination of the order-parameter amplitude at the superconductor-insulator transition in 2D. Physical Review Letters, 69(24), 3567.

130. Vélez, M., Martín, J. I., Villegas, J. E., Hoffmann, A., González, E. M., Vicent, J. L., & Schuller, I. K. (2008). Superconducting vortex pinning with artificial magnetic nanostructures. Journal of Magnetism and Magnetic Materials, 320(21), 2547–2562.

131. Wang, Q. H., & Lee, D. H. (2003). Quasiparticle scattering interference in high-temperature superconductors. Physical Review B, 67(2), 020511.

132. Weygand, D., Bréchet, Y., & Lépinoux, J. (1999). Zener pinning and grain growth: a two-dimensional vertex computer simulation. Acta Materialia, 47(3), 961–970.

133. Xing, W., Shi, P., Song, H., Zhao, Y., & Li, L. (2019). Global pinning synchronization of stochastic delayed complex networks. Information Sciences, 490, 113–125.

134. Xu, S. Y., Li, Q., Wertz, E., Hu, Y. F., Pogrebnyakov, A. V., Zeng, X. H., & Redwing, J. M. (2003). High critical current density and vortex pinning of epitaxial MgB 2 thin films. Physical Review B, 68(22), 224501.

135. Yao, C., & Ma, Y. (2021). Superconducting materials: Challenges and opportunities for large-scale applications. Iscience, 24(6), 102541.

136. Yoon, T. H., & Cho, M. (2021). Quantitative complementarity of wave-particle duality. Science Advances, 7(34), eabi9268.

137. Yuan, P., Xu, Z., Wang, D., Zhang, M., Li, J., & Ma, Y. (2016). Vortex pinning properties in Co-doped $BaFe_2As_2$ thin films with a high critical current density over 2 MA cm^{-2} at 9 T. Superconductor Science and Technology, 30(2), 025001.

138. Zampoli, V., & Jordan, P. M. (2022). Second-sound phenomena in type II conductors with Stefan–Boltzmann source. Mechanics Research Communications, 126, 103998.

CHAPTER

3 Chemistry and High Temperature Superconductors

Contents

3.1. INTRODUCTION

Superconductors have zero electrical resistance and behave as perfect diamagnets (known as the Meissner effect). This arises from the condensation of electrons near the Fermi level into Cooper pairs that behave as a collective quantum mechanical state – a superconductor is a charged superfluid. Thermal pair-breaking limits superconductivity to a maximum critical 25 temperature (T_c) above which the material shows metallic or semiconducting behavior with a finite resistance. Superconductivity is also limited by critical magnetic fields (H_c) and current densities (J_c) at temperatures below T_c. The critical temperature is mainly determined by chemical composition and structure, whereas the critical fields and currents are also strongly influenced by microstructure and are often not optimum in homogenous materials (Attfield, 2011).

The electron-pairing interactions are relatively weak, e.g., in comparison to magnetic exchange interactions, so that all known superconductors have T_c's well below ambient temperature. Cryogenic cooling is thus needed to exploit the useful properties of superconductors such as in power transmission cables, magnetically levitated trains, or SQUID (superconducting quantum interference device) electronics based on the Josephson effect of tunnelling Cooper pairs. Increasing T_c towards ambient temperature is thus a major ambition for superconductivity research (Krabbes et al., 2006).

The first century of superconducting materials research divides into two eras, the 'low-T_c' period from the 1911 discovery of the zero-electrical resistance transition in mercury until 1986 when the record critical temperature was T_c = 23 K in Nb_3Ge, and the subsequent 'high-T_c' era, during which several types of chemically complex solids with T_c's up to 138 K have emerged (Haskel et al., 2001).

This review will briefly describe the seven distinct families of high-T_c superconductors (taken to be those with $T_c \geq 23$ K at ambient pressure) that have been discovered in the last quarter century. The families vary in size from containing one to many chemically and structurally similar materials with a common physical mechanism for superconductivity. The overall chemical trends and possible future directions for materials discovery are also discussed.

3.2. HIGH-T_C FAMILIES

3.2.1. Cuprates

High (critical) temperature superconductivity was born from the discovery of an unprecedented T_c = 35 K transition in Ba-doped La_2CuO_4 in 1986. A flurry of activity in the following years led to the identification of many more superconducting cuprates, with the highest T_c of 138 K found in $HgBa_2Ca_2Cu_3O_{8+\delta}$. A fluorinated sample of this phase showed an onset T_c of 166 ± 1 K 20 at 23 GPa pressure which is the highest measured superconducting critical temperature to date, and approaches the lowest recorded terrestrial temperature of 184 K. An enormous literature is available for the cuprate superconductors; some recent books and review articles are cited here. Although the pairing mechanism and a convincing explanation for the magnitude of T_c in this family remain controversial, the essential chemical features are clearly established (Chu et al., 2015).

Tip

Understanding the different families of high-temperature superconductors, such as cuprates, fullerides, and iron arsenides, allows for a deeper exploration of their unique chemical properties.

The complex chemistry of the cuprates results from the requirement for several structural features, to optimize superconductivity:

1. Copper oxide planes are essential. These have stoichiometry CuO_2 and a geometry like that found in the MO_2 planes of the AMO_3 perovskite structure. Hence cuprates are sometimes described as 'layered perovskites.' Maximum T_c's are generally found for materials with blocks of three adjacent, hole-doped CuO_2 planes.
2. Electropositive cations, usually from the alkaline earth (Ca, Sr, Ba) or rare earth (La-Lu, Y) metals, act as layer separators in two distinctive structural roles. Large 'A' cations (typically Sr^{2+}, Ba^{2+}, or La^{3+}) support additional coordination of a further oxygen to copper, and this can provide a connection to additional metal (M) oxide layers. Smaller electropositive 'B' cations (usually Ca^{2+} or a small rare earth) separate CuO_2 planes in multilayer cuprates without allowing intercalation of O between Cu's in adjacent planes which is detrimental to superconductivity. n CuO_2 planes require (n-1) B cation spacer layers.

3. Blocks of one or two covalent metal oxide layers MO_x (M can be Cu, Ru, Hg, Tl, Pb, Bi) layers are connected to CuO_2 planes via interplanar oxides in the AO layers. The MO_x layers are sometimes termed the 'charge reservoir' as they compensate for the doping of the CuO_2 planes although this can also achieved by non-aliovalent substitutions at the A or B sites (Anderson et al., 2004).

Hence many cuprates have compositions $(MO_x)_m(AO)_2B_{n-1}(CuO_2)_n$ = $M_mA_2B_{n-1}Cu_nO_z$, often abbreviated as M–m,2(n–1),n, e.g., the highest-T_c material $HgBa_2Ca_2Cu_3O_{8+\delta}$ is abbreviated as Hg-1223.

The undoped cuprates contain Cu^{2+} within the CuO_2 sheets and are antiferromagnetic insulators, with very strong antiferromagnetic Cu-O-Cu super exchange interactions of coupling strength J/k ~ 1500 K. Superconductivity is induced by doping the CuO_2 sheets; electron doping (reduction of Cu^{2+}) is effective in a few cases, e.g., Nd_2CuO_4, but for the majority including all the highest-T_c materials, hole-doping is achieved by cation substitutions, increasing oxygen content, or by band overlap (Fradkin et al., 2015). The latter two mechanisms are found in many materials, e.g., maximum T_cs ≈ 40 K are obtained for x = 0.16 and y = 0.08 in the $La_{2-x}Sr_xCuO_4$ and La_2CuO_{4+y} systems respectively, both of which yield an average +2.16 oxidation state for Cu.

Hole-doping through band overlap is less common, but is well-established in the important 93 K superconductor $YBa_2Cu_3O_7$ (also known as YBCO or (Y)123). This contains distinct CuO chains as the MO_x layers in the above structural classification and Cu band overlap results in a formal charge distribution $Cu^{2.6+}Ba_2YCu^{2.2+}{}_2O_8$. Similar Ru-Cu charge transfer induces superconductivity in the ruthenocuprate $Ru^{4.8+}Sr_2GdCu^{2.1+}{}_2O_8$ (Rybicki et al., 2016).

A simplified electronic phase diagram for the hole-doped cuprates. Initial doping of a Cu^{+2} parent material disrupts long range antiferromagnetic order and the Néel transition is suppressed at 3% doping. Superconductivity appears above 5% doping – the intermediate region is found to be physically in homogenous with both magnetic and superconducting correlations present. Further oxidation increases T_c to a maximum at 15–20% doping, above which superconductivity is suppressed and is no longer apparent above ~25% 20 hole doping. There is substantial evidence for a diffuse high temperature insulator-metal transition at which a pseudo-gap opens, and this coincides with T_c for the over doped superconductors. Above T_c, the cuprates have unusual normal state electronic properties that evidence strong electron-electron correlations, but at high doping levels they become more like x) (Figure 3.1).

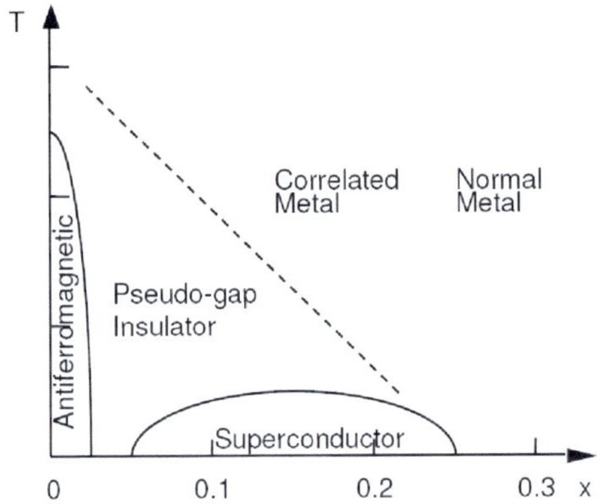

Figure 3.1. *Schematic electronic phase diagram for cuprate superconductors as a function of the hole doping x (equivalent to average Cu oxidation state 2+x).*

Source: Barišić et al.; Creative Commons License.

Remember

Identifying and studying the chemical commonalities among various high-temperature superconductors provides valuable insights into the underlying mechanisms and potential pathways for discovering new superconducting materials.

In addition to the doping level, structural features are also very important for optimizing T_c in the cuprates. Buckling of the Cu-O-Cu bridges in the CuO_2 sheets suppresses superconductivity and favors alternative charge and spin ordered insulating ground states, so large Ba^{2+} cations at the A type cation sites help to preserve flat CuO_2 planes and high T_c's. Disorder arising from mixed A or B cations adjacent to the planes also suppresses superconductivity. Hence the highest reported T_c for single layer cuprates is 98 K for optimally doped $HgBa_2CuO_{4+\delta}$ ($\delta = 0.08$) which has only Ba^{+2} cations adjacent to the CuO_2 planes, and relatively little additional strain and disorder from the small concentration δ of oxygen interstitials between the Hg sites. Coupling between nearby CuO_2 layers separated by B cation layers also enhances T_c; the highest values are found in the Hg-family where T_c increases up to 138 K for n = 3. T_c decreases for higher n most probably because the doping becomes non-uniform across 15 inequivalent

CuO_2 planes. The pairing mechanism for superconductivity in the cuprates remains unclear. The essential features for theoretical descriptions are the d-wave symmetry of the order parameter (the wavefunction describing the Cooper pairs has the symmetry of a $d_{x^2-y^2}$ atomic orbital), the presence of strong antiferromagnetic correlations, and the pseudogap feature. A plausible explanation is that antiferromagnetic fluctuations mediate the pairing instead of the electron-phonon coupling found in conventional BCS (Bardeen, Cooper and Schrieffer) type materials (Kivelson et al., 2003).

3.2.2. Fullerides

The discovery of high temperature superconductivity in copper oxides was followed by another remarkable finding from a very different group of materials. After buckminsterfullerene (C_{60}) was first identified and isolated in the 1980's, alkali metal fulleride derivatives were synthesized and superconductivity was first reported in K_3C_{60} with T_c = 19 K. The fulleride superconductors are highly air-sensitive which hinders their characterization and limits their practical utility. Nevertheless, the variation of superconductivity across the A_3C_{60} family has been explored in detail and related alkaline earth and lanthanide-doped fullerides, with T_c's up to 8 K for Ca_5C_{60} and $Sm_{2.75}C_{60}$, have also been prepared (Murphy et al., 1991).

The main tuning parameter for the A_3C_{60} superconductors is the interfulleride separation, represented by the unit cell volume per C_{60}. T_c increases up to a maximum of 33 K for $RbCs_2C_{60}$ as the volume increases, but beyond this limit superconductivity is destabilised with respect to an antiferromagnetic magnetic ground state. Recent studies have found that T_c increases under pressure up to 35 and 38 K in the face- and body-centered cubic polymorphs of Cs_3C_{60}, respectively.

The phonon-mediated BCS mechanism for superconductivity in the weak coupling limit describes many aspects of the fullerides. Vibrational spectra reveal a coupling between the conduction electrons and high frequency vibrations of the C_{60}^{3-} anions that strengthens as the interfulleride separation increases, and a BCS-type ^{13}C isotope effect is also observed. However, the transition to a Mott (magnetic) insulating state at limiting high separations is more similar to the breakdown of metallic and superconducting behavior at low dopings in cuprates and other unconventional high-T_c superconductors (Bill & Kresin, 2002).

3.2.3. Barium Bismuthate

The perovskite $BaBiO_3$ undergoes charge disproportionation that results in a distorted crystal structure containing an ordered alternation of Bi^{3+} and Bi^{5+} sites. Suppression of charge order in $Ba_2Bi^{3+}Bi^{5+}O_6$ to give a superconductor was first demonstrated by substituting Pb for 5 Bi, and the maximum T_c in this system is 13 K. Renewed interest following the discovery of superconductivity in cuprates led to the further discovery of T_c's up to 30 K in $Ba_{1-x}K_xBiO_3$. The phase diagram for this system, superconductivity emerges with the maximum T_c immediately beyond the suppression of the charge ordered state at x = 0.38 and diminishes with further doping up to the x = 0.5 stability limit of the solid solution. As superconductivity is only found in the three-dimensional bismuthate perovskite structure there are few chemical variations. $Sr_{1-x}K_xBiO_3$ and $Sr_{1-x}Rb_xBiO_3$ analogues have been prepared at high pressure, but these have lower T_c's of 12 and 13 K respectively. The large radii and good size matching of Ba^{2+} and K^+ ions appear to be optimal for superconductivity in doped-20 $BaBiO_3$ (Norton & Tang, 1991).

Optical spectra show that $Ba_{1-x}K_xBiO_3$ is an s-wave superconductor (like conventional low-T_c materials), but the electron-phonon coupling constant was found to be too small for conventional BCS coupling to explain the high T_c. Electron-phonon coupling may be enhanced by electron-electron interactions in negative-(Hubbard) U models, which follow the negative-U description of the disproportionation of the average s configuration to give ordered s and s states in the $BaBiO_3$ parent material (Sutto et al., 1991).

3.2.4. Quaternary Borocarbides

Many rare earth transition metal borides and carbides are superconducting but most have low T_c values. However, a specific family of quaternary materials with composition RM_2B_2C (R = rare earth, M = Ni, Pd) were discovered to have T_c's up to the past record value of 23 K for YPd_2B_2C. The structure consists of layers of isolated linear B_2C groups between R and M layers. Magnetic R cations suppress superconductivity completely for R = Pr, Nd, Sm, Gd, and Tb in the RNi_2B_2C series and a variety of antiferromagnetic states is found, but coexistence of superconductivity and magnetic order is observed for the later R = Dy, Ho, Er, 10 Tm. This strong coupling of the rare earth metal moments to the conduction electrons is in contrast to the $RBa_2Cu_3O_7$ and RFeAsO series where the R-magnetism has little influence on superconductivity. The RM_2B_2C superconductors appear to be s-wave materials but with an anisotropic energy gap (Mazumdar & Nagarajan, 2015) (Figure 3.2).

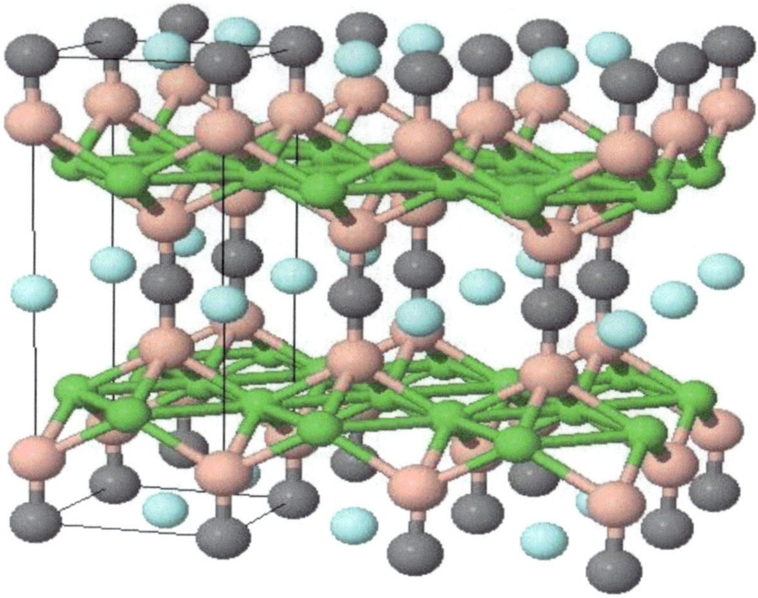

Figure 3.2. *Crystal structure of the YPd_2B_2C superconductor showing metal-metal bonded Pd layers connected by linear BCB units, with Y atoms in the C plane.*

3.2.5. Intercalated Nitride Halides

MNX (M = Zr, Hf; X = Cl, Br, I) phases are insulators that contain hexagonal $X(MN)_2X$ layers which may be stacked in several polymorphic arrangements. Chemical or electrochemical intercalation of alkali metals (Li, Na, K) into the van der Waals gaps between the layers or removal of a small amount of halogen X, dopes electrons into the M d-band inducing superconductivity. The maximum observed T_c in this family is in the original report of 25.5 K for $Li_{0.48}(THF)_yHfNCl$ containing intercalated tetrahydrofuran solvent (THF). Intercalation staging has been observed in Na_xHfNCl, and T_c's of 24 and 20 K were reported for the stage 1 and 2 materials respectively. The intercalated MNX phases are very air-sensitive which has hampered study of their physical properties (Yamanaka, 2010).

Superconductivity with maximum T_c = 16.3 K has recently been reported in A_xTiNCl (A = Li, Na, K, Rb), where the constituent layers have an orthorhombic (FeOCl type) structure that differs from that of the Zr and Hf based materials. This reveals that superconductivity is not specific to one underlying lattice symmetry in this family (Fuertes, 2014). Physical measurements show that the nitride halides are not conventional BCS superconductors and have a large superconducting gap ratio $2\Delta/k_BT_c = 4.6–5.6$. In the Li_xZrNCl system, the maximum T_c is observed for minimum $x = 0.06$ doping, below which a magnetic insulating state is observed (Zhang & Xu, 2022).

3.3. MAGNESIUM DIBORIDE

The discovery of high-T_c superconductivity at 39 K in MgB_2 in 2001 was one of the most unexpected developments in this field. Transition metal diborides were explored extensively during the low-T_c era but the magnesium analogue, which is a standard, air-stable chemical reagent, was thought to be uninteresting as it is has no available d-states. Subsequent measurements have shown that MgB_2 can have a T_c of up to 41 K in thin films and can be processed into practical conductors, with critical current densities of up to $J_c = 3.4 \times 10^7$ Acm^{-2} reported (Collings et al., 2008).

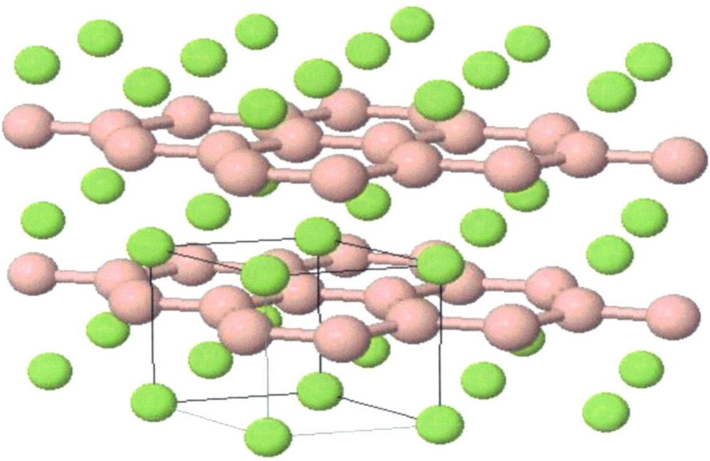

Figure 3.3. *The hexagonal structure of the MgB_2 superconductor showing graphitic boron sheets interleaved by layers of magnesium.*

Source: Royal Society of Chemistry, Creative Commons License.

MgB_2 has a layered structure with magnesium atoms between graphitic boron sheets. It appears to be an optimum superconductor 'as is,' and doping or substitutions of other metals for Mg or of C for B have not enhanced T_c. However, this discovery has inspired reinvestigation of graphite intercalation compounds resulting in several new superconductors with T_c's up to 11.5 K in CaC_6. Extensive physical measurements have shown that MgB_2 is a near-perfect BCS superconductor, with an isotope-effect increase of approximately 1 K in T_c when B is replaced by B. An important feature is the presence of two superconducting gaps with $2\Delta/k_BT_c$ values of 1.1 and 4.0 resulting from Cooper pairing of electrons in σ and π-bands, respectively. The latter value is close to the BCS s-wave value of 3.53. Coupling between the two gap pairings results in the single, observed superconducting transition (Monteverde et al., 2001) (Figure 3.3).

3.4. IRON ARSENIDES

The final family provides an appropriately symmetric closure to the first quarter century of high temperature superconductor discovery, with very high T_c's second only to those of the cuprates, and many chemical and physical similarities. They are based on FeAs layers in which Fe is tetrahedrally coordinated by As atoms and several structure types with different additional layers are known. A comprehensive review of the field has recently been published. High-T_c's were first reported in the electron-doped $LaFeAsO_{1-x}F_x$ series and subsequent study of rare earth $RFeAsO_{1-x}F_x$ and oxygen-deficient $RFeAsO_{1-\delta}$ analogues led to discovery of the highest T_c = 55 K to date in $SmFeAsO_{1-x}F_x$. Hole-doped materials are also superconducting; $La_{0.85}Sr_{0.15}FeAsO$ has T_c = 25 K, and in the related AFe_2As_2 and AFeAs families, $Ba_{0.6}K_{0.4}Fe_2As_2$ has T_c = 38 K and LiFeAs has T_c = 18 K (Sebastian et al., 2008).

Superconductivity is also observed in analogues where Fe or As are replaced by similar elements, but with lower T_c's. The binary phase $Fe_{1+x}Se$ containing only the FeAs-type layers is superconducting with T_c = 9 K for a small Fe excess x = 0.01, and T_c increases up to 14 K for $Fe_{1+x}Se_{0.6}Te_{0.4}$. LaFePO has T_c = 6.6 K. Non-Fe analogs such as LaNiPO and LaNiAsO, BaM_2As_2 (M = Cr, Mn, Co, Ni, Ru, Rh) and BaM_2P_2 (M = Ni, Rh, Ir), and LiFeP have T_c's <5 K (Kuchinskii & Sadovskii, 2009).

The electronic phase diagram for the high-T_c iron arsenide superconductors appears similar to that of the cuprates, although an important difference is that the parent materials are 15 metallic whereas the undoped cuprates are Mott insulators. Hence the observed antiferromagnetic order of small (0.1–1 μ_B) Fe moments is probably a spin density wave rather than an array of local moments. Doping suppresses the long range antiferromagnetism and superconductivity emerges with a high T_c for 10–20% doping. The competing ground states stabilise different lattice symmetries. The spin order is antiferromagnetic along one of the two in-plane axes but ferromagnetic in the perpendicular direction, which drives an orthorhombic distortion of the ideal tetragonal structure. However, superconducting phases are tetragonal as the spin density wave is suppressed. The structural transition occurs above the Néel temperature, and probably reflects the onset of antiferromagnetic fluctuations. Coexisting antiferromagnetic and superconducting phases are found in the crossover region for some systems. T_c is very sensitive to lattice effects and is optimized when the As-Fe-As angle in the tetrahedral layers is close to the geometric ideal of 109.5°(Qi et al., 2008).

Two gaps with s-wave symmetry have been found in many experiments and the values of $2\Delta/k_BT_c$ fall into the range 1–4 and 5–9. As for MgB_2, coupling between the two orders results in a single superconducting transition. The emergence of superconductivity from a consistent magnetic state in the iron arsenides provides strong evidence for an antiferromagnetic spin fluctuation pairing model, as for the cuprates (Norman, 2008) (Figure 3.4).

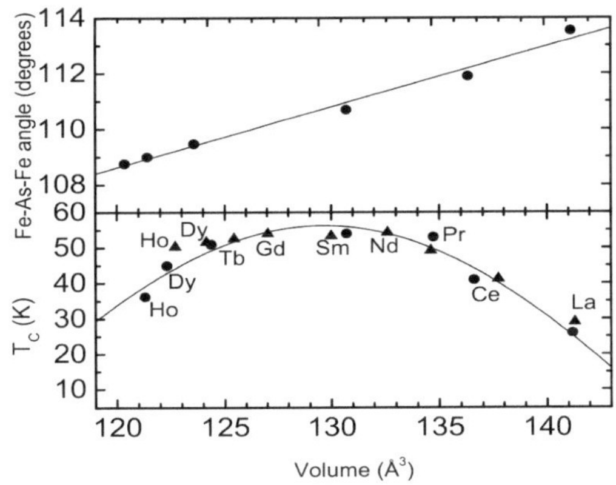

Figure 3.4. *Variations of Fe-As-Fe angle (upper panel) and the maximum reported T_c (lower panel) with unit cell volume for optimally doped $RFeAsO_{1-x}F_x$ (circles) and $RFeAsO_{1-\delta}$ (triangles) superconductors, with rare earths R as shown. The overall maximum T_c = 56 K obtains for an angle of 110.6°, close to the ideal tetrahedral angle.*

Source: Si et al.; Creative Commons License.

3.5. CHEMICAL COMMONALITIES

The seven families of high temperature superconducting materials do not fall into a welldefined chemical group. However, one general feature that clearly distinguishes the high-T_c set from the metals and alloys that dominated the low-T_c era is apparent from the simple chemical sorting. This follows the standard classification of the elements as metals or nonmetals. A plot of T_c against the atomic fraction of nonmetal shows that all of the highest T_c 20 materials have at least 50% nonmetal content, whereas Nb_3Ge and the alloys that dominated the low-T_c era have <50% nonmetal. This plot follows the Edwards and Sienko classification of elements based on refractivity/volume ratios, in which As and Ge are respectively a nonmetal and metal, but the above observation is unchanged if Ge is taken to be a nonmetal. In only one of the seven families, the iron arsenides, are some high-T_c materials with <50% nonmetal content observed, and even here the highest-T_c member of the family, $SmFeAsO_{1-x}F_x$, has 50% nonmetal content. In this context, the AFe_2As_2 family with T_c's up to 38 K in $Ba_{0.6}K_{0.4}Fe_2As_2$ are notable as the only metal-rich materials to have exceeded the 23 K limit (Korb, 2018).

A well-known physical truism is that '*a good superconductor is a bad metal*,' meaning that the electron-pairing interactions that give rise to superconductivity also diminish conductivity in the normal state above T_c. The equivalent chemical statement is thus that '*a good superconductor is mostly nonmetal*,' meaning that the narrow bands and strong electronelectron correlations required for high temperature superconductivity are found in conducting 10 materials with a high nonmetal content (Young et al., 2010).

As for many metal-nonmetal compounds, the electron distribution in high-T_c superconductors and their parent materials can be represented to a first approximation by ionic formulae based on typical valence states, e.g., $(La^{3+})_2Cu^{2+}(O^{2-})_4$ and $Sm^{3+}Fe^{2+}As^{3-}O^{2-}$, although the distribution $Y^{3+}(Pd_2^{3+})(B_2C^{6-})$ implies some Pd-Pd bonding as is observed in the quaternary borocarbide crystal structure. Further consideration of the roles of the metal and nonmetal in the high-T_c families shows that there are two limiting cases. The first is where only metal to nonmetal bonding is significant. These 'metal-nonmetal' families (cuprates, iron arsenides, bismuthates and nitride halides) have structures that follow simple ionic principles, with metal and nonmetal atoms bonded to each other but not to themselves, leading to near 50:50 metal:nonmetal 25 compositions (between 40:60 and 60:40 for the high-T_c materials (Shemetulskis et al., 1996). The oxide and nitride halide superconductors have average anion charges of –2 and the presence of higher valent cations leads to nonmetal contents >50%. The AFe_2As_2 (and AFeAs) arsenides fall below the 50% limit because the magnitude of the anion charge is greater than that of the cations.It is apparent that the highest-T_c members of the cuprates and iron arsenides are those with compositions closest to 50% nonmetal content, but the significance of this observation is unclear. An important chemical feature is that a specific metal-nonmetal pair is required to generate high-T_c's in these families. No obvious analogues to the Cu-O and Bi-O combinations have been reported, but in the other two cases optimum pairings with $T_c > 23$ K/suboptimal pairings with $T_c < 23$ K are evident; Fe-As/Fe-P, Ni-P, etc., for iron pnictides; Hf-N/Zr-N,Ti-N for nitride halides. This demonstrates that superconductivity is very sensitive to the degree of nonmetal to metal charge transfer that occurs through orbital hybridisation. This group includes the most unconventional (least BCS-like) families, where metal-nonmetal orbital hybridisation is also important in strengthening the magnetic exchange that may mediate spin fluctuation superconductivity in cuprates and iron arsenides (Tran et al., 2000).

In the second 'nonmetal-bonded' limit, nonmetal to nonmetal bonding is important, and the cations play a subsidiary role. This is best exemplified by the A_3C_{60} fullerides where the alkali metal cations act only as spacers for the fulleride anions and do not contribute to the electronic states near the Fermi level. MgB_2 is close to this limit although here some B to Mg charge transfer is important to the electronic structure. The presence of covalent bonding between 20 nonmetal atoms leads to a lower anion charge per atom and hence higher nonmetal contents; MgB_2 contains $(B^-)\infty$ sheets and the fullerides contain discrete C_{60}^{3-} anions. This group of high temperature superconductors is more BCS-like, with nonmetal-nonmetal bonding leading to narrow bands and high vibrational frequencies, and hence strong electron-phonon coupling. However, the antiferromagnetism observed at the limit of superconductivity in fullerides may 25 also be relevant (Sulmon et al., 2015).

The quaternary borocarbides are intermediate between the above two limits as they contain strongly bonded, discrete B_2C groups but also show chemical specificity for Pd over Ni or other metals and have a 50:50 metal-nonmetal composition. Of the seven high temperature superconductor families, this is the only one to show prominent metal-metal bonding, and as the 30 T_c does not exceed that of Nb_3Ge, it could equally be regarded as 'low T_c' (Comfort et al., 2007).

3.6. FUTURE PROSPECTS

The first 25 years of the high-T_c era have been spectacularly productive, with a new $T_c \geq 23$ K superconductor family discovered every few years. As noted above, each family is chemically distinct from the others, so prediction of future discoveries is difficult, however some indicators are evident. Based on the known high-T_c materials, future high-T_c superconductors are likely to have a substantial (≥50%) nonmetal content. Very high T_c's are associated with a specific pairing in the 'metal-nonmetal' group, and have nonmetal contents of 40–60% and structures in keeping with simple ionic bonding considerations. Neither element may have featured in previous highT_c families, as for Fe and As prior to 2008 (Flores-Livas et al., 2020). Spin-spin fluctuations appear to be important to the superconducting mechanism in the highest-T_c cuprates and iron arsenide materials where magnetic transition metals are needed.

However, f-block magnetism may also be beneficial, as found in the intermetallic heavy fermion superconductor $PuCoGa_5$ which has $T_c = 18$ K, so incorporation of nonmetals into such materials might prove a useful aid to future discoveries. Non-magnetic mechanisms based on other electronic instabilities may also emerge, as for the bismuthate superconductors based on suppression of charge disproportionation. Chemical doping is required to suppress the spin or charge ordered ground state and induce superconductivity in the metal-nonmetal families. This is typically achieved through nonaliovalent substitutions or non-stoichiometry, but can sometimes occur through a fortuitous band overlap, as in $YBa_2Cu_3O_7$. However, disorder within the essential metal-nonmetal network tends to suppress superconductivity, and so additional parts of the structure that can be chemically tuned (the 'charge reservoir' in cuprates) are needed to obtain high-T_c's (Namburi et al., 2021).

Did You Know?

Ongoing research and exploration of high-temperature superconductors hold promise for uncovering new materials and expanding our understanding of the complex chemistry involved, paving the way for future advancements in superconductivity.

This is illustrated by the difference between the maximum T_c's of 13 K for $BaBi_{1-x}Pb_xO_3$ doped at the essential Bi sites, and 30 K for $Ba_{1-x}K_xBiO_3$ doped at the secondary Ba sites. The second group of high-T_c materials is characterized by nonmetal-nonmetal bonding which may lead to very high nonmetal contents (potentially 100% for a purely organic superconductor). This group is limited to those elements that form strongly bonded covalent molecules and networks, typically B and C, although similar nonmetals such as N, O, Si, P, S could also be incorporated. Superconductivity is

BCS-like in this group and the maximum observed T_c to date of 41 K is in keeping with optimal BCS weak-coupling predictions. In these families the optimal electronic structures for superconductivity are achieved for the stoichiometric compositions A_3C_{60} and MgB_2 (and also YPd_2B_2C) and chemical doping does not raise T_c (Feng et al., 2015).

Structurally, most of the high-T_c families are based on layered arrangements, although the cubic structures of A_3C_{60} and $Ba_{1-x}K_xBiO_3$ show that this is not a strict requirement. Twodimensionality may be physically important for enhancing the pairing fluctuations needed for superconductivity. However, layered structures also offer far greater chemical and structural 35 flexibility than three-dimensional networks, and so it is more likely that the optimal conditions for superconductivity can be realized. The highest-T_c $HgBa_2Ca_2Cu_3O_{8+\delta}$ material represents one of approximately 30 chemically or structurally distinct cuprate subfamilies. Predictions of the highest-possible T_c's given the present known materials are extremely difficult. As estimates based on physical understanding of the mechanisms for high temperature superconductivity are still limited, a statistical approach may be the most realistic approach to gauge the likelihood of discovering new high-T_c materials. Extreme value theory describes the probabilities of substantial deviations from the median in a large collection of values generated from a given set of rules, and is used to model the likelihood of occurrence of rare events such as mechanical failures or extreme weather. Three limiting distributions are found, of which the 10 Fréchet type is appropriate to a variable bounded by a lower limit such as $T_c > 0$ here. The Fréchet probability distribution function is (Zurek & Bi, 2019):

$$f(T_c) = (\alpha/\beta)(\beta/T_c)^{1+\alpha}\exp(-(\beta/T_c)^{\alpha})$$

and the cumulative distribution function is:

$$F(T_c \leq T) = \exp(-(\beta/T)^{\alpha})$$

The distribution of maximum T_c values from the seven families known to have maximum $T_c \geq 23$ K (as shown in Table 1) plus Nb_3Ge. Assuming that the total number of presently known superconductor families is $N_{Tot} \sim 1000$, $f(T_c)$ values were obtained in 10 K intervals as n/N_{Tot} where n is the number of superconductor families with maximum T_c in the ranges $23 \leq T_c/K < 33$, $33 \leq T_c/K < 43$, etc. The $f(T_c)$ fit shown has parameters $\alpha = 1.5$ and $\beta = 5$. Improved estimates of these parameters could be obtained by analyzing maximum T_c values 25 in the low-T_c region (Figure 3.5).

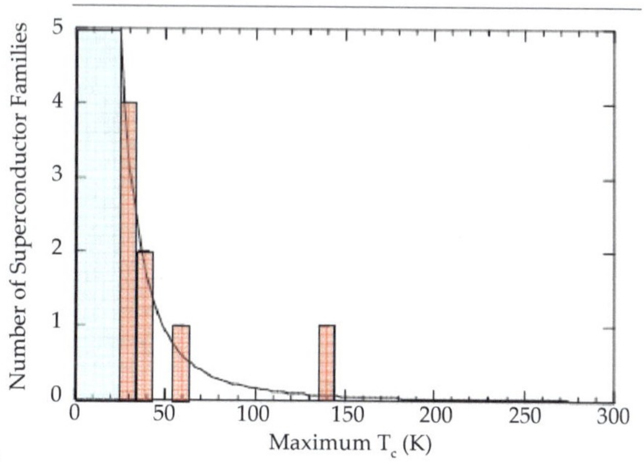

Figure 3.5. *The distribution of superconductor families according to their maximum T_c, binned in 10 K intervals for $T_c \geq 23$ K materials. The curve has equation $1000f(T_c)$ with Fréchet function parameters as shown in the text.*

Source: Sato et al.; Creative Commons License.

The probability of a superconductor family having maximum $T_c > T$ is $(1–F(T_c \leq T))$ so from the above parameter values, the probability of a newly discovered superconductor family having maximum $T_c > 100$ K is 1%, decreasing to 0.2% for $T_c > 300$ K (room temperature superconductivity). These statistical estimates appear up to an order of magnitude too high as only one presently known family has $T_c > 100$ K, so ranges of 0.1–1% and 0.02–0.2% may be more realistic, and even these should be treated with appropriate caution. On a statistical basis, several hundred new families of superconductor may have to be discovered to find another $T_c > 100$ K material, and perhaps many hundreds or a few thousand to realize the possibility of superconductivity at room temperature. However, more targeted approaches based on knowledge of known materials as presented here, or perhaps on new theoretical insights, should reduce the chemical space to be explored (McDevitt et al., 1993).

Activity 3.1

You are a chemist working in a laboratory specializing in high-temperature superconductors. Develop a hands-on experiment that demonstrates the influence of different chemical compositions and their impact on the critical temperature of superconducting materials.

3.7. SUMMARY

Chapter three focuses on the chemistry of high-temperature superconductors. It begins with an introduction to the chapter's topics and then explores different families of high-TC (critical temperature) superconductors. This includes an examination of cuprates, fullerides, barium bismuthate, quaternary borocarbides, and intercalated nitride halides. The chapter also discusses specific materials such as magnesium diboride and iron arsenides, highlighting their relevance in high-temperature superconductivity. Furthermore, the chapter examines the chemical commonalities among various high-temperature superconductors, allowing for a better understanding of the underlying chemical mechanisms and potential pathways for discovering new superconducting materials. It concludes with a look towards future prospects in the field, indicating the ongoing research and potential advancements in high-temperature superconductivity.

REVIEW QUESTIONS

1. What is the significance of understanding the chemistry of high-temperature superconductors?
2. Describe the cuprate family of high-temperature superconductors.
3. What are fullerides and their relevance in superconductivity?
4. Discuss the characteristics and properties of barium bismuthate as a high-temperature superconductor.
5. What are quaternary borocarbides and their role in superconductivity?
6. Explain the concept of intercalated nitride halides as high-temperature superconductors.
7. What is the importance of magnesium diboride in the field of superconductivity?
8. Describe the chemistry and properties of iron arsenides as high-temperature superconductors.
9. How do chemical commonalities among high-temperature superconductors contribute to our understanding of the field?
10. What insights can be gained from identifying chemical commonalities among different families of high-temperature superconductors?

REFERENCES

1. Anderson, P. W., Lee, P. A., Randeria, M., Rice, T. M., Trivedi, N., & Zhang, F. C. (2004). The physics behind high-temperature superconducting cuprates: the 'plain vanilla' version of RVB. *Journal of Physics: Condensed Matter*, *16*(24), R755.
2. Attfield, J. P. (2011). Chemistry and high temperature superconductivity. *Journal of Materials Chemistry*, *21*(13), 4756–4764.
3. Barišić, N., Chan, M. K., Li, Y., Yu, G., Zhao, X., Dressel, M., & Greven, M. (2013). Universal sheet resistance and revised phase diagram of the cuprate high-temperature superconductors. *Proceedings of the National Academy of Sciences*, *110*(30), 12235–12240.
4. Bill, A., & Kresin, V. Z. (2002). Intercalation and high temperature superconductivity of fullerides. *The European Physical Journal B-Condensed Matter and Complex Systems*, *26*, 3–6.
5. Chu, C. W., Deng, L. Z., & Lv, B. (2015). Hole-doped cuprate high temperature superconductors. *Physica C: Superconductivity and Its Applications*, *514*, 290–313.
6. Collings, E. W., Sumption, M. D., Bhatia, M., Susner, M. A., & Bohnenstiehl, S. D. (2008). Prospects for improving the intrinsic and extrinsic properties of magnesium diboride superconducting strands. *Superconductor Science and Technology*, *21*(10), 103001.
7. Comfort, D. A., Bobrov, K. S., Ivanen, D. R., Shabalin, K. A., Harris, J. M., Kulminskaya, A. A., & Kelly, R. M. (2007). Biochemical analysis of Thermotoga maritima GH36 α-galactosidase (Tm GalA) confirms the mechanistic commonality of clan GH-D glycoside hydrolases. *Biochemistry*, *46*(11), 3319–3330.
8. Feng, X., Zhang, J., Gao, G., Liu, H., & Wang, H. (2015). Compressed sodalite-like MgH 6 as a potential high-temperature superconductor. *RSC Advances*, *5*(73), 59292–59296.
9. Flores-Livas, J. A., Boeri, L., Sanna, A., Profeta, G., Arita, R., & Eremets, M. (2020). A perspective on conventional high-temperature superconductors at high pressure: Methods and materials. *Physics Reports*, *856*, 1–78.
10. Fradkin, E., Kivelson, S. A., & Tranquada, J. M. (2015). Colloquium: Theory of intertwined orders in high temperature superconductors. *Reviews of Modern Physics*, *87*(2), 457.
11. Fuertes, A. (2014). Chemistry and superconductivity of intercalated metal nitride halides. *Semiconductor Science and Technology*, *29*(6), 064005.
12. Haskel, D., Stern, E., Dogan, F., & Moodenbaugh, A. R. (2001). Dopant

structural distortions in high-temperature superconductors: an active or a passive role? *Journal of Synchrotron Radiation*, *8*(2), 186–190.

13. Kivelson, S. A., Bindloss, I. P., Fradkin, E., Oganesyan, V., Tranquada, J. M., Kapitulnik, A., & Howald, C. (2003). How to detect fluctuating stripes in the high-temperature superconductors. *Reviews of Modern Physics*, *75*(4), 1201.
14. Korb, J. (2018). Chemical fertility signaling in termites: idiosyncrasies and commonalities in comparison with ants. *Journal of Chemical Ecology*, *44*(9), 818–826.
15. Krabbes, G., Fuchs, G., Canders, W. R., May, H., & Palka, R. (2006). *High Temperature Superconductor Bulk Materials.* Weinheim: Wiley-Vch. (Vol. 1, pp. 250–252).
16. Kuchinskii, E. Z., & Sadovskii, M. V. (2009). Multiple bands: a key to high-temperature superconductivity in iron arsenides? *JETP letters*, *89*, 156–160.
17. Mazumdar, C., & Nagarajan, R. (2015). Quaternary borocarbides: Relatively high Tc intermetallic superconductors and magnetic superconductors. *Physica C: Superconductivity and Its Applications*, *514*, 173–183.
18. McDevitt, J. T., Riley, D. R., & Haupt, S. G. (1993). Electrochemistry of high-temperature superconductors. Challenges and opportunities. *Analytical Chemistry*, *65*(11), 535A-545A.
19. Monteverde, M., Nunez-Regueiro, M., Rogado, N., Regan, K. A., Hayward, M. A., He, T., & Cava, R. J. (2001). Pressure dependence of the superconducting transition temperature of magnesium diboride. *Science*, *292*(5514), 75–77.
20. Murphy, D. W., Rosseinsky, M. J., Haddon, R. C., Ramirez, A. P., Hebard, A. F., Tycko, R., & Dabbagh, G. (1991). Superconductivity in alkali metal fullerides. *Physica C: Superconductivity*, *185*, 403–408.
21. Namburi, D. K., Shi, Y., & Cardwell, D. A. (2021). The processing and properties of bulk (RE) BCO high temperature superconductors: current status and future perspectives. *Superconductor Science and Technology*, *34*(5), 053002.
22. Norman, M. R. (2008). High-temperature superconductivity in the iron pnictides. *Physics*, *1*, 21.
23. Norton, M. L., & Tang, H. Y. (1991). Superconductivity at 32 K in electrocrystallized barium potassium bismuth oxide. *Chemistry of Materials*, *3*(3), 431–434.

24. Qi, Y., Gao, Z., Wang, L., Wang, D., Zhang, X., & Ma, Y. (2008). Superconductivity at 34.7 K in the iron arsenide Eu_0.7Na0.3Fe_2As_2. *New Journal of Physics*, *10*(12), 123003.
25. Rybicki, D., Jurkutat, M., Reichardt, S., Kapusta, C., & Haase, J. (2016). Perspective on the phase diagram of cuprate high-temperature superconductors. *Nature Communications*, *7*(1), 11413.
26. Sato, K. I., Kobayashi, S. I., & Nakashima, T. (2011). Present status and future perspective of bismuth-based high-temperature superconducting wires realizing application systems. *Japanese Journal of Applied Physics*, *51*(1R), 010006.
27. Sebastian, S. E., Gillett, J., Harrison, N., Lau, P. H. C., Singh, D. J., Mielke, C. H., & Lonzarich, G. G. (2008). Quantum oscillations in the parent magnetic phase of an iron arsenide high temperature superconductor. *Journal of Physics: Condensed Matter*, *20*(42), 422203.
28. Shemetulskis, N. E., Weininger, D., Blankley, C. J., Yang, J. J., & Humblet, C. (1996). Stigmata: an algorithm to determine structural commonalities in diverse datasets. *Journal of Chemical Information and Computer Sciences*, *36*(4), 862–871.
29. Si, Q., Yu, R., & Abrahams, E. (2016). High-temperature superconductivity in iron pnictides and chalcogenides. *Nature Reviews Materials*, *1*(4), 1–15.
30. Sulmon, C., Van Baaren, J., Cabello-Hurtado, F., Gouesbet, G., Hennion, F., Mony, C., & Gérard, C. (2015). Abiotic stressors and stress responses: What commonalities appear between species across biological organization levels? *Environmental Pollution*, *202*, 66–77.
31. Sutto, T. E., Averill, B. A., & Ruvalds, J. (1991). Superconductivity in barium bismuth lead thallium oxide ($Ba_4BiPb_2TlO_{12-\delta}$). *Chemistry of Materials*, *3*(2), 209–211.
32. Tran, N. L., Locke, P. A., & Burke, T. A. (2000). Chemical and radiation environmental risk management: differences, commonalities, and challenges. *Risk Analysis*, *20*(2), 163–172.
33. Yamanaka, S. (2010). Intercalation and superconductivity in ternary layer structured metal nitride halides (MNX: M= Ti, Zr, Hf; X= Cl, Br, I). *Journal of Materials Chemistry*, *20*(15), 2922–2933.
34. Young, S. L., Wilson, M. J., Hillier, S., Delbos, E., Ali, S. M., & Stoltzfus, R. J. (2010). Differences and commonalities in physical, chemical and mineralogical properties of Zanzibari geophagic soils. *Journal of Chemical Ecology*, *36*, 129–140.
35. Zhang, Y., & Xu, X. (2022). Predicting the superconducting transition temperature of high-Temperature layered superconductors via machine

learning. *Physica C: Superconductivity and Its Applications*, *595*, 1354031.

36. Zurek, E., & Bi, T. (2019). High-temperature superconductivity in alkaline and rare earth polyhydrides at high pressure: A theoretical perspective. *The Journal of Chemical Physics*, *150*(5), 050901.

CHAPTER

4 Materials Aspects of High-Temperature Superconductors for Applications

Contents

4.1. INTRODUCTION

Cuprate high-temperature superconductors (HTS) continue to be a prominent area of focus in the field of physics, even 16 years after their initial discovery. The mechanism behind superconductivity remains a topic of active discussion within the academic community. The explanation of their typical state characteristics has proven to pose a greater obstacle to the theoretical framework of Solid-State Physics. The advancement in elucidating these inquiries has been significantly attributed to the enhancement of sample quality, as evidenced by sources.

The emergence of numerous applications envisioned during the scientific excitement following the discovery of high-temperature superconductivity can be partially attributed to the endeavors of Applying Superconductivity in establishing a materials foundation. The current state of HTS material research bears a resemblance to the historical progression of semiconductor materials advancement from 50 years ago. Contrary to silicon, high-temperature superconductors (HTS) are complex multi-element compounds that rely on intricate sequences of oxide layers, as evidenced by various studies. Apart from the challenge posed by the presence of extraneous elements, which hindered the development of consistent material properties in the early stages of semiconductor research, high-temperature superconductors (HTS) also encounter inherent defects in local stoichiometry resulting from the incorporation of cations in an incorrect layer as well as defects within the oxygen sublattice. In order to optimize the superconducting characteristics, it is necessary to modify the amount of oxygen according to a compound-specific off-stoichiometric ratio, as a further requirement. However, it is important to ensure that the leading oxygen vacant spaces or interstitials are distributed in a spatially uniform microscopic manner. This has been previously noted in the literature (Bednorz & Müller, 1986).

The emergence of HTS posed a formidable challenge to the scientific community in the study of a group of intricate compounds that would have otherwise been encountered through a traditional research approach involving a gradual escalation of material complexities, a process that would have taken a considerable amount of time to achieve. The pursuit of enhanced quality in HTS materials has been impeded by the limited superconducting uniformity length, which is comparable to the dimensions found in a crystallographic section cell, thereby posing a significant challenge to advancements in this field. Conversely, the imposition of experimental limitations establishes a strong correlation between superconducting properties that can be measured on a large scale as well as microstructural characteristics, thereby enabling the surveillance of the performance of HTS materials at anatomical levels. The extensive endeavors in high-throughput screening (HS) materials science have facilitated the advancement of the enhanced synthesis of oxide materials in a broad sense, such as ferroelectric in nature oxides or just manganites. The diverse range of atypical physical characteristics exhibited by this latest cohort of "advanced oxides" has introduced novel opportunities for both fundamental and practical investigations (Orenstein & Millis, 2000).

Currently, there exist reproducible methods for preparing various HS materials, which serve as a fundamental basis for their practical applications. Fortuitously, the enhancement of the superconducting properties of these materials appears to align with endeavors to enhance their stability within technical settings, despite the fact that these substances are only metastable chemically under such circumstances (Anderson, 1987).

4.2. ATOMIC STRUCTURE AND CLASSIFICATION

The arrangement of mobile battery carriers within HTS substances is determined by a structural component consisting of stacked CuO_2 layers, with a variable number of n = 1, 2, 3... These layers are interconnected through intermediate Ca sections, as reported in previous studies. The "active blocks" of $(CuO_2/Ca/)n{-}1CuO_2$ stacks have a corresponding counterpart known as "charge reservoir blocks" EO/(AOx)m/EO, where m = 1, 2 monolayers containing a random oxide AOx (such as Bi, Pb, Tl, Hg, Au, Cu, Ca, B, Al, Ga) are enclosed on every side through a monolayer for alkaline earth oxide EO with E = Ba, Sr. The arrangement of HTS is a consequence of the alternating arrangement of the two previously mentioned block units. The selection of BaO or SrO as the "wrapping" layered is not a random decision, but rather is contingent upon the specific AOx involved, as it must facilitate a favorable spatial alignment between the CuO_2 and AO_x layers (Anderson, 1997) (Figure 4.1).

The chemical equation denoted as $A{-}m_2(n{-}1)_n$, where A represents a combination of $A_mE_2Ca_{n-1}Cu_nO_{2n+m+2+y}$, is a commonly used abbreviation. For instance, $Bi_2Sr_2Ca_2Cu_3O_{10}$ is abbreviated as Bi-2223, wherein the alkaline earth component is not explicitly indicated (refer to Tab.1). The set of representatives for all values of n, ranging from 1 to infinity, that share an identifiable AOx will be denoted as "A-HTS" in this context. For instance, an example of A-HTS would be Bi-HTS (Figure 4.2).

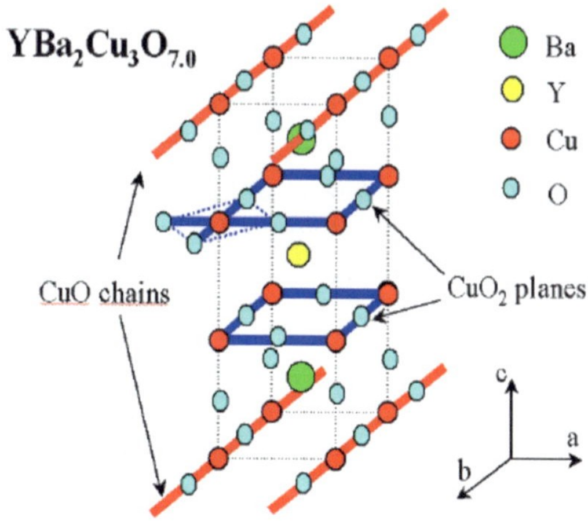

Figure 4.1. *Illustration of the crystal structure of* $YBa_2Cu_3O_7$ *("YBCO"). The presence of the CuO chains introduces an orthorhombic distortion of the unit cell (a = 0.382 nm, b = 0.389 nm, and c = 1.167 nm).*

Source: Kinoshita, Creative Commons License.

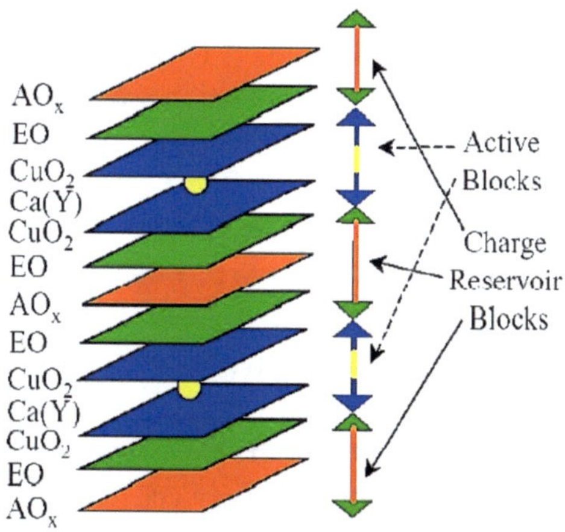

Figure 4.2. *General structure of a cuprate HTS* $A\text{-}m_2(n-1)_n$ $(A_mE_2Ca_{n-1}Cu_nO_{2n+m+2+y})$ *for m = 1. For m = 0 (or 2) the missing (additional) AOx layer per unit cell leads to a (a/2, b/2, 0) "side step" of the unit cells adjoining in c-axis direction.*

Source: Kinoshita, Creative Commons License.

The compound $YBa_2Cu_3O_7$, is the most notable example of a high-temperature superconductor. It was the first of its kind to be discovered, using a temperature threshold (Tc) for the beginning of superconductivity that exceeds the temperature at which it boils of liquid nitrogen. This compound is commonly referred to as "YBCO" or "Y-123" ($Y_1Ba_2Cu_3O_{7-\delta}$) in scientific literature. The material in question conforms to the broader HTS classification system as a variant of Cu-1212, wherein Y fully replaces Ca. The observed phenomenon can be attributed to the disparity in valence between Y (+3) and Ca (+2), resulting in the generation of additional charge adjacent to the CuO_2 layers.

The replacement scheme used in the HTS substances $RBa_2Cu_3O_{7-\delta}$ ("RBCO," "R-123"), wherein R may substitute La or any other rare earth component with the exception of Ce or Tb, can be considered as a broadening of this scheme.

The utilization of the lanthanide contractions of the R electrons presents an empirical means of determining the separation distance within the two CuO_2 layers within the operative block of the infused Cu-1212 mix. $Y_1Ba_2Cu_4O_8$, also known as Y-124, is considered an m = 2 equivalent of Cu-2212 in the YBCO system. The HTS substances denoted as "214" with examples such as $La_{2-x}Sr_xCuO_4$ ("LSCO") or $Nd_{2-x}Ce_xCuO_4$ ("NCCO"), are somewhat unconventional in this classification system. However, they can also be expressed as "0210" using the values of m = 0, n = 1, and E_2 residing $La_{2-x}Sr_x$ and $Nd_{2-x}Ce_x$, correspondingly (Batlogg et al., 1994).

Additional noteworthy chemical alterations of the fundamental HTS compositions involve the incorporation associated with fluorine or just chlorine as substituents of oxygen that become more electronegative. Frequently, this leads to a minor enhancement of Tc. The superconductor Hg-1223 exhibits a transition temperature (Tc) of 135 K, which has been reported to increase to 138 K.

This represents the highest Tc observed thus far under normal pressure conditions. However, under conditions of elevated pressure of 30 GPa, a Tc of 164 K. Oxycarbonate high-temperature superconductors (HTS) have been implemented in numerous HTS compounds. The remarkable aspect of incorporating carbon through the structure of high-temperature superconductors (HTS) is that it has an impartial or even slightly advantageous impact on their superconducting characteristics, as evidenced by previous research (Batlogg, 1998).

This is noteworthy because carbon impurities, when present as secondary states in HTS samples, typically have a highly detrimental impact on their superconducting properties (Table 4.1).

Table 4.1. Tabular Representation of the Present Study Pertains to the Categorization and Documented Critical Temperature (T_c) Values of High-Temperature Superconducting (HTS) Compounds

HTS Family	Stoichiometry	**Notation**	**Compounds**	**Highest T_c**
Bi-HTS	$Bi_mSr_2Ca_{n-1}Cu_nO_{2n+m+2}$ m = 1, 2 n = 1, 2, 3…	Bi-$m_2(n-1)_n$, BSCCO	Bi-1212	102 K
			Bi-2201	34 K
			Bi-2212	90 K
			Bi-2223	110 K
			Bi-2234	110 K
Pb-HTS	$Pb_mSr_2Ca_{n-1}Cu_nO_{2n+m+2}$	Pb-$m_2(n-1)_n$	Pb-1212	70 K
			Pb-1223	122 K
Tl-HTS	$Tl_mBa_2Ca_{n-1}Cu_nO_{2n+m+2}$ m = 1, 2 n = 1, 2, 3…	Tl-$m_2(n-1)_n$, TBCCO	Tl-1201	50 K
			Tl-1212	82 K
			Tl-1223	133 K
			Tl-1234	127 K
			Tl-2201	90 K
			Tl-2212	110 K
			Tl-2223	128 K
			Tl-2234	119 K
Hg-HTS	$Hg_mBa_2Ca_{n-1}Cu_nO_{2n+m+2}$ m = 1, 2 n = 1, 2, 3…	Hg-$m_2(n-1)_n$, HBCCO	Hg-1201	97 K
			Hg-1212	128 K
			Hg-1223	135 K
			Hg-1234	127 K
			Hg-1245	110 K
			Hg-1256	107 K
			Hg-2212	44 K
			Hg-2223	45 K
			Hg-2234	114 K
Au-HTS	AumBa2Can-1CunO2n+m+2	Au-m2(n–1)$_n$	Au-1212	82 K

123-HTS	RBa2Cu3O7-δ R = Y, La, Pr, Nd, Sm, Eu, Gd, Tb, Dy, Ho, Er, Tm, Yb, Lu	R-123, RBCO	Y-123, YBCO Nd-123, NBCO Gd-123 Er-123 Yb-123	92 K 96 K 94 K 92 K 89 K
Cu-HTS	CumBa2Can-1CunO2n+m+2 m = 1, 2 n = 1, 2, 3…	Cu-m2(n-1)n	Cu-1223 Cu-1234 Cu-2223 Cu-2234 Cu-2245	60 K 117 K 67 K 113 K < 110 K
Ru-HTS	$RuSr_2GdCu_2O_8$	Ru-1212	Ru-1212	72 K
B-HTS	BmSr2Can-1CunO2n+m+2	B-m2(n-1) n	B-1223 B-1234 B-1245	75 K 110 K 85 K
214-HTS	E_2CuO_4	LSCO “0201”	La2-xSrxCuO4 Sr2CuO4	51 K 25 (75) K
		Electron-Doped HTS PCCO NCCO	La2-xCexCuO4 Pr2-xCexCuO4 Nd2-xCexCuO4 Sm2-xCexCuO4 Eu2-xCexCuO4	28 K 24 K 24 K 22 K 23 K
	Ba2Can-1CunO2n+2	“$02(n-1)_n$”	“0212” “0223” “0234” “0245”	90 K 120 K 105 K 90 K
Infinite-Layer HTS	$ECuO_2$	*Electron-Doped I.L.*	Sr1-xLaxCuO2	43 K

Source: Timmer, John, Creative Commons License.

4.3. STRUCTURAL OBSTACLES FOR SUPERCURRENTS

Following the initial excitement surrounding High-Tc superconductivity in 1987, a subsequent period of disappointment ensued upon the discovery of critical current density measurements. The most recent density values (Jc) achieved in the most superior polycrystalline material accessible during that period were only a few hundred A/cm2. These values were lower than the prevailing density achieved in copper wiring at ambient temperature, which is associated with minimal energy dissipation. The circumstances were exacerbated by the fact that even modest magnetic fields were found to be capable of extinguishing these suboptimal supercurrents. The current state of LTS materials does not appear to align with its previous success story, as the largest market share is now dominated by magnet applications, which serve as the foundation for a significant wire industry. The phenomenon of "weak-link behavior" in high-temperature superconductors (HTS) has been extensively studied and can be effectively mitigated through the implementation of specific processing methodologies (Erb et al., 1997).

Tip

Understanding the structural obstacles that can hinder the flow of supercurrents is crucial for optimizing the performance of high-temperature superconductors in practical applications.

The phenomenon whereby superconductivity is inhibited by magnetic fields generated by electrical currents is well-documented. For a considerable duration, the utilization of superconductors in high-current applications was impeded. The initial stride towards achieving this objective involved the identification of type-II superconductors, wherein the magnetic permeability (λ) is significantly greater than the superconductive coherence length (ξ). This characteristics enables the magnetic fields to infiltrate the superconducting bulk. The simultaneous presence of magnetic radiation as well as superconductivity results in a significant decrease in the dissipation of energy associated with the condensation of superconducting particles, which is necessary for the absorption of magnetic fields. This phenomenon allows for the persistence of superconductive properties regardless of the presence of high-intensity magnetic fields. Luckily, high-temperature superconductors (HTS) exhibit characteristics of extreme type-II superconductivity, with a penetration depth (λ) greater than 100 nanometers and a coherence length (ξ) of approximately 1 nanometer (Yang et al., 1999).

The phenomenon of superconductivity in high-temperature superconductors (HTS) is thought to stem from the physics associated with the CuO_2 layers, which house the mobile charges. The interlayer linking of the CuO_2 layers in a $(CuO_2/Ca/)n\text{-}1CuO_2$ stack exhibits weaker superconductivity compared to the intralayer linking across the CuO2 layers. However, it is still considerably stronger compared to the Josephson linking that exists between $(CuO_2/Ca/)n\text{-}1CuO_2$ piles. The quasi-two-dimensional characteristics of superconductivity in high-temperature superconductors (HTS) results in a significant anisotropy of their superconducting properties. This anisotropy is manifested by substantially higher supercurrents throughout the CuO_2 layers compared to the direction that is perpendicular to it. Although this property remains poorly understood in terms of practical applications, it might be mitigated through supplementary engineering endeavors (Yamamoto et al., 2001) (Figures 4.3 and 4.4).

Remember

Thin films of high-temperature superconducting materials offer unique advantages, such as flexibility and compatibility with microfabrication techniques, making them suitable for various technological applications.

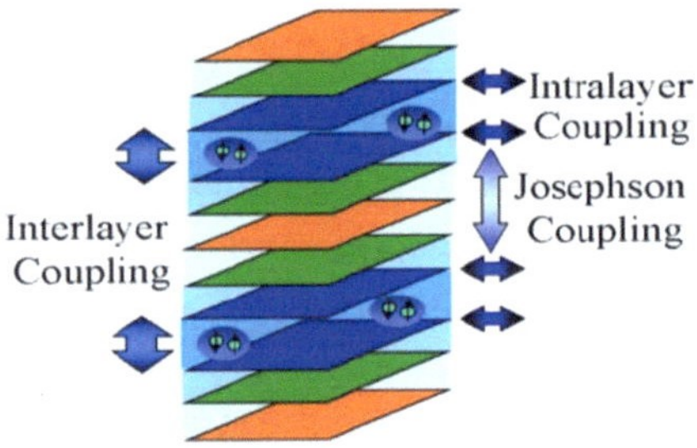

Figure 4.3. *Illustration of hierarchy of the superconducting.*

Source: Aidam, Creative Commons License.

The coupling among distinct entities into pancake vortices is attributed to the presence of feeble superstructural components in cuprate high-temperature superconductors. The investigation pertains to the analysis of interlayer coupling within ducting and the schematic overlap for neighboring vortices.

Merely possessing the characteristics of being a type-2 superconductor does not suffice for facilitating the unimpeded flow of potent currents without experiencing dissipation. The introduction of magnetic swirls into the superconductive material through the magnetic field necessitates their fixation by pinning centers, lest the flux movement through the vortices precipitate dissipation. In technological superconductors, the coherent length's material flaws impede superconductivity within these areas, creating a natural location for vortex centers to reside (Service, 1996).

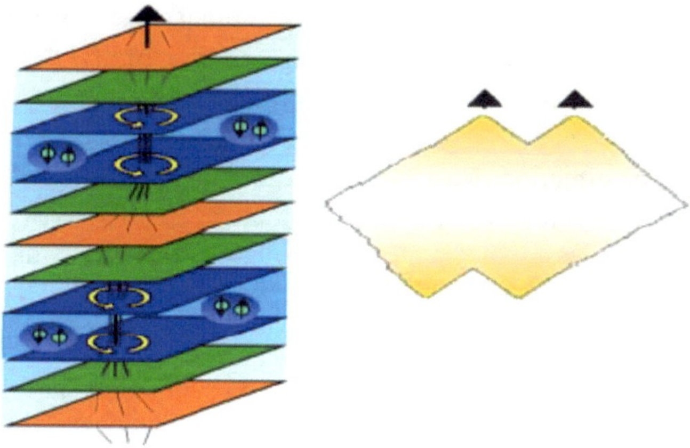

Figure 4.4. *Quasi-disintegration of magnetic vortex lines.*

Source: Aidam, Creative Commons License.

The small consistency lengths of HTS, such as YBCO with coherence length measurements of ξab = 1.6 nm and ξc = 0.3 nm over T → 0 K [66], make them susceptible to encountering material imperfections that affect the consistency length dimensions. These coherence length measures are already analogous to the structure parameters used in YBCO, those are a = 0.382 nm, b = 0.389 nm, and c = 1.167 nm. Nonetheless, the low critical superconducting coupling (ξc) between the (CuO2/Ca/)n-1CuO2 piles gives rise to novel challenges. The thickness associated with the charge storage container blocks, which are situated between the piles of EO/(AOx)m/EO, exceeds ξc. Consequently, the vortices in this region become ill-defined because of the sparse distribution of Cooper pairs (Johnson et al., 1993).

The phenomenon results in a partial degeneration of the vortices, forming stacks of "pancake vortices" that exhibit greater flexibility compared to the quasi-rigid vortex-like lines found in typical superconductors. As a consequence, these pancake vortices necessitate individual tying centers. The degree of quasidisintegration varies among HTS compounds due to the fact that the value of ξc is comparable to the degree of thickness of just one oxide layer. The total number of layers present in the charged storage blocks for EO/(AOx)m/EO has a notable impact on the pinning characteristics as well as consequent supercurrents within magnetic fields. YBCO, also known as "Cu-1212," exhibits a superior supercurrent capacity within magnetic fields compared to Bi-HTS Bi-2212 and Bi-2223, which remain the most prevalent HTS conductor content due to production-related factors (Chu, 1999).

In addition to the inherent challenges associated with the transportation of supercurrent within single-crystalline HTS substances, there exist further impediments due to the fact that these substances are typically not a uniform continuum, instead being a complex network of interconnected grains. The process of crystal growth entails the displacement of excess material that cannot be accommodated within the lattice framework of the developing grains. This displaced material is subsequently pushed towards the growth front, leading to the accumulation of residual secondary states as well as contaminants at the interfaces among the grains. The aforementioned obstacles hinder the present transportation system and must be circumvented through meticulous regulation of the growth procedure, specifically with regard to the structure of the provided substance (Chu, 1997) (Figure 4.5).

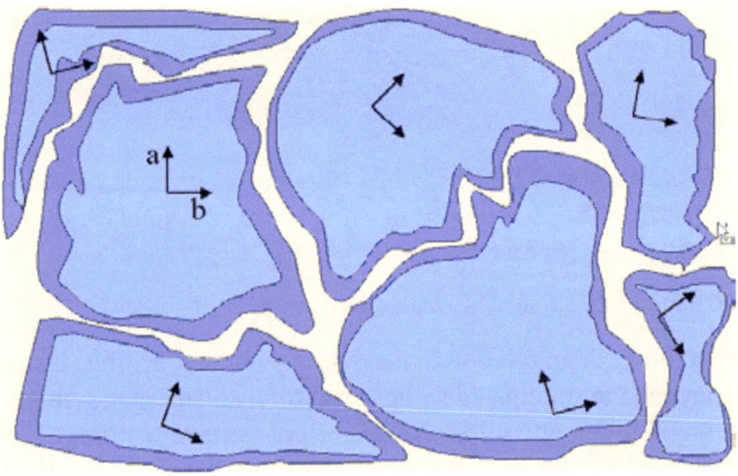

Figure 4.5. *Schematics of the HTS microstructure: Differently oriented single crystal grains are separated by regions filled with secondary phases. In addition, oxygen depletion may occur at grain boundaries.*

Source: Roland Hott, Creative Commons License.

Misalignment for grains poses a significant obstacle over supercurrents in high-temperature superconductors (HTS). An exponential decline in the movement of supercurrent is noted with respect to the misalignment position. The established d-symmetry that defines the superconducting sequence parameter within the majority of high-temperature superconducting substances is considered a contributing factor to this action (Figure 4.6).

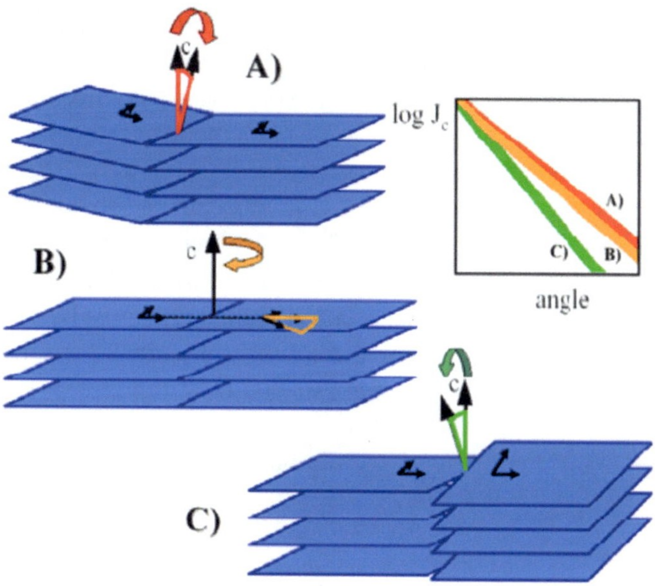

Figure 4.6. *Illustration of basic grain boundary geometries and experimentally observed* J_c *reduction* $J_c \sim e^{-\alpha/\alpha_0}$ *as function of the misalignment angle α:* $\alpha_0 \approx 5°$ *for (A) and (B),* $\alpha_0 \approx 3°$ *for (C) independent of temperature.*

Source: Roland Hott, Creative Commons License.

The decline of Jc in relation to the disarray angle α exhibits a significant increase beyond the anticipated value based solely on d-wave symmetry. The degradation of Jc and alterations in the current-voltage features of transport action have been attributed to structural flaws, including dislocations along with deviations coming from stoichiometry. These defects are thought to be the cause of the observed changes. The reduction of oxygen present at the outermost layers of grains results in a decline in doping in comparison to the mass value within the grain. Consequently, there is a localized deterioration of the superconducting characteristics in accordance with the temperature-doping schematic of high-temperature superconductors. A recent study has shown that Ca-doping can mitigate this phenomenon to some extent in

YBCO, thereby expanding the permissible range regarding grain disarray in high-temperature superconducting materials (Tallon et al., 2000). To sum up, the production of high-temperature superconducting materials that can be practically utilized necessitates precise preparation methods to achieve chemically pure grains with minimal misalignment of their crystalline axes.

4.4. TECHNICALLY APPLICABLE HTS MATERIALS

The advancement of high-temperature superconducting (HTS) materials with practical applications has been pursued through various avenues. Epitaxial thin sheets of high-temperature superconductors exhibit remarkable superconducting characteristics, including a critical temperature (Tc) exceeding 90°K and a critical current density below 500 micro-ohms. Additionally, the films demonstrate a high value of the resistivity-temperature characteristics (Rs(T,f)) and a significant current density (Jc) at 77°K and zero magnetic fields (f = 0) that surpasses 2.

These properties render the films highly appropriate for application in superconductive technology. Microwave filtering techniques utilized in commercial as well as military applications currently employ a microwave surface resistivity Rs (77 K, 10 GHz) of 106 A/cm2. Josephson junctions utilizing Josephson tunneling via ultra-thin artificial obstacles or just the Josephson junction-induced behavior of high-temperature superconductor (HTS) grain borders were developed and can be employed in the fabrication of extremely sensitive electromagnetic sensors, commonly known as Superconducting Quantum Interference Devices (SQUIDs). In addition, active electronic components are evaluated to determine their potential to expand the scope of high-temperature superconducting thin-film apps. The application potential of HTS singular crystals is limited due to their diminutive size as well as low Jc numbers, which can be attributed to a deficient density of tying centers. Nevertheless, the melt-textured high-temperature superconducting (HTS) bulk material, despite being a simplified version, exhibits exceptional magnetic pinning characteristics and is currently utilized as high-field constant magnets in various applications like magnetic drives as well as motors. Despite the ceramic properties inherent in cuprate oxides, the production of flexible high-temperature superconductor (HTS) wire or adhesive conductor substance is achieved through either the incorporation of HTS as slender filaments within a silver-based matrix or via the application of an HTS coating onto metallic carrier tapes (Liitgemeier et al., 1996).

The Bi-HTS conductors material stands out as the sole anomaly within the category of high-tech HTS materials, as it does not require a robust biaxial texture to attain currents that are of practical significance. The Bi-22(n-1) n HTS exhibits remarkable softness due to the feeble Van-der-Waals connection between the adjacent BiOx layers within the charging reservoir layers, thereby facilitating alignment through mechanical processing techniques such as rolling or just pressing. An additional factor contributing to this phenomenon is the favorable mechanical as well as electrical interconnection within the Ag matrix as well as the surrounding materials, facilitated by their closely matched melting temperatures. This enables the passage of high levels of current even in situations where direct current transmission within the HTS grains fails, as a slight deviation through the Ag medium is feasible. The aforementioned statement is also relevant to the Bi-2212 bulk that has been impregnated with Ag and is readily obtainable in various sizes exceeding 10 cm, exhibiting a consistent Jc (77 K, 0 T) of multiple kA/cm^2 (Ahn et al., 1995; Salamon & Jaime, 2001).

Polycrystalline HTS substances have demonstrated significantly elevated critical current density values. However, their implementation entails a greater degree of risk in practical applications. Whenever there is a significant decrease in Jc in comparison to Jc (77 K, 0 T) ~ 1 MA/cm^2 observed in well-textured substances with an adequate number of pinning centers, it suggests the presence of considerable structural and electrical heterogeneity. This condition can have severe consequences. The imposition of current by means of areas of faulty materials has the potential to induce a localized loss of superconductivity along with the emergence of "hot spots." The energy released during quenching induces structural and chemical alterations in the vicinity of the "hot spots" within the high-temperature superconductor material. The aforementioned alterations have the potential to compromise the ideal circumstances for the efficient transfer of supercurrent between adjacent high-temperature superconducting grains, thereby resulting in an expansion of the quench region. The propagation of the quench region could potentially result in the degradation of the superconducting material. The quenching phenomenon observed in this scenario differs from that observed within classical superconductors, wherein the quenching is attributed to inadequate heat transfer resulting from the phonon system freezing at low temperatures of liquid helium. At elevated operating temperatures of high-temperature superconductors (HTS), the problem of spreading heat through phonons becomes less critical due to its highly effective nature, resulting in a significantly elevated specific heat (Sun, 2001).

4.5. THIN FILMS

Currently, YBCO as well as Tl-HTS (Tl-2212) lightweight films are being manufactured in significant quantities, numbering in the thousands of wafers per annum, for the purpose of commercial as well as military microwave filtering methods. The investigation of RE-123 films (where RE represents elements that include Nd, Sm, Gd, Dy, etc.) as potential alternatives to YBCO has been conducted, with findings indicating slightly elevated Tc values (Moreo et al., 1999). Nevertheless, the elevated processing temperature that is necessary can augment the possibility of fissures in the films. The softness of Bi-HTS thin films renders them unsuitable for various applications. The fiber of cotton has been observed to effectively remove high-quality Bi-HTS films. The production of Hg-HTS films necessitates the use of sealed quartz tubes to achieve high-quality and stable films, thereby limiting their application to smaller substrate sizes (Renner et al., 2002).

The commercial accumulation technique for YBCO has been created as thermal energy evaporation. Substrates of varying sizes, with a maximum diameter of 9 inches and an area of 20 x 20 cm^2, may be coated upon a single side at a standard rate of 20–30 nm/min. The bilateral coating may be achieved through a consecutive pair of deposition procedures. In the meantime, it is worth noting that pulsed laser deposition (PLD) exhibits a degree of proximity to these aforementioned possibilities. The utilization of magnetic sputtering with deposition rates reaching 10 nm/min is a prevalent practice in various research facilities (Mawdsley et al., 1992; Brosha et al., 1993). Preparations are currently underway for the extension of bigger wafer sizes. The implantation of vast areas of DyBCO using molecular beam epitaxy (MBE) was effectively demonstrated in recent times. However, the process is characterized by intrinsic disadvantages such as lower deposit rates as well as the more complicated design of the depositing machines. The MOCVD technique has demonstrated the capability of achieving depositing rates of up to 1 μm/min. However, it has been observed that the deposits of YBCO films with high-quality characteristics have been observed at significantly lower rates. The stability associated with the Ba precursor seems a significant issue. The sol-gel method has shown significant advancements in the manufacture of YBCO lightweight films, particularly in terms of cost-effectiveness. This is due to the ability to rapidly deposit precursor films without the need for vacuum conditions, followed by annealing to create high-quality epitaxial films (Yamauchi et al., 1996). The elevated critical temperature (Tc) exceeding 100 K exhibited by Tl-2212 enables a greater capacity for operational temperatures in comparison to YBCO.

The standard technique for fabricating involves precursor deposition and subsequent annealing within Tl_2O_3 vapor, owing to the elevated volatility for Tl (Yamauchi & Karppinen, 2000). In contemporary times, there has been a notable enhancement in microwave characteristics. Nevertheless, this has come at the cost of comparable inclinations towards the emergence of cracks with the rise in film dimension, akin to the case of YBCO.

In order to achieve uniform orientation of HTS grains, it is necessary for substrates to furnish a crystal matrix that is appropriately lattice matched. The c-axis introductions of HTS films, wherein the CuO_2 layers lie parallel with the substrate, are deemed most appropriate for various applications as it maintains the isotropy for electrical characteristics within the film plane. The expeditious development of HTS films that follow the CuO_2 plane orientations in contrast with the c-axis direction serves as a propitious factor in the advancement of this film orientation. Ensuring that the heating coefficients for the substrates along with the HTS films are matched is an essential prerequisite, given that the HTS depositing process is carried out within a temperature range of 650–900°C. The mechanical pressure resulting from the differential contraction caused by the HTS film as well as the substrate upon cooling to ambient or cryogenic temperatures is a limiting factor over the film's dimension, beyond which crack formation occurs (Bordet et al., 1997). The depth of YBCO films upon silicon along with sapphire substrates are restricted to approximately 50 nm as well as 250–300 nm, correspondingly, which poses a limitation. In addition, it is necessary to buffer such substrates using very thin coatings of CeO_2 as well as Y-stabilized ZrO2 (YSZ) so as to avoid the detrimental effect of Si or Al dissemination through the HTS films, which can lead to the suppression of superconductivity. Buffering layers offer advantages in terms of enhanced lattice matching.

The technique of wet etching has the potential to be employed for the purpose of creating patterns on device structures that have dimensions as small as those in the micrometer range. The occurrence of uneven edges in the photoresist stencils presents a challenge in achieving uniform sidewalls (Wu et al., 1999). Consequently, the technique of electron beam engraving is the favored approach for producing structures of micrometer scale (Wu et al., 1987). By utilizing e-beam lithography as well as supplementary preparation methods, it is feasible to fabricate structures alongside sizes below 1 μm (Lin et al., 1995). Although HTS single-layer networks have achieved commercial reproducibility, the production of multilayer networks remains limited to low yield. The multiple-layer technology, which requires high-temperature accumulation steps, is exclusively applicable to YBCO owing to the extremely volatile nature of Tl, as reported in references (Chu et al., 1992; Williams et al., 1996). Advancements in insulating properties as well as elevated current capacity within various YBCO layers, coupled with improved integrability regarding Josephson junctions, provide optimizm for the enhanced implementation of high-temperature superconductor (HTS) circuits with greater complexity.

4.6. JOSEPHSON JUNCTIONS

The electrical transportation action of HTS junctions is significantly more sensitive than that of classical superconductor intersections due to the d-wave balance of the superconductive wave function as well as the proximity concerning the metal-insulator change. The emergence of Attached Andreev states as well as the consequent reduction of the ordered parameter at interactions, which is contingent upon their orientation relative to the crystalline axes, pose supplementary physical challenges in terms of ensuring spatial uniformity of Jc. The selection of YBCO as the preferred material for HTS junctions constructed by Josephson is attributed to its compatibility with multilayer structures (Ai-Mamouri et al., 1994) (Figure 4.7).

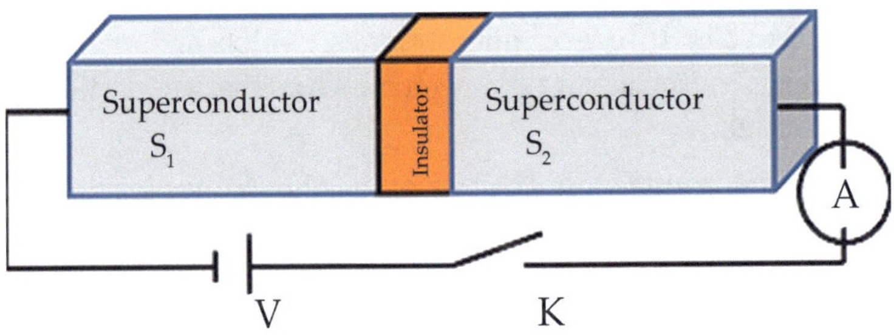

Figure 4.7. *Schematic of Josephson junction.*

Source: Md. Rafiqul Islam, Creative Commons License.

Did You Know?

Bi-2212 is a well-known high-temperature superconducting material used in wire and tape conductors, known for its high critical temperature and critical current density.

According to current research, ramp-type intersections remain the most viable junction form for high-temperature superconducting digital circuits.

A certain level of replication of contact variables has been achieved to a degree that enables the fabrication of digital networks that have a complexity ranging from 10 to 100 junctions. Previous barrier concepts have been developed using thin insulation layers made of (Ga-doped) PrBCO or Co-doped YBCO. However, the homogeneity regarding the critical current density (Jc) was found to be restricted by the growth regulation of the protective layer with the ramp.

The article discusses the implementation of "interface-engineered" intersections, wherein a "natural barrier" (most likely "pseudo-cubic" YBCO) is created through Ar ion imprint attitudes of the ramp-based electrode, as opposed to the use of a deposited barrier layer. Studies have reported enhanced reproducibility along with stability in successive extreme temperature procedures for the preparation of multi-layer materials (Kawashima et al., 1996; Hiroi et al., 1994). Regrettably, it has been observed that barrier characteristics exhibit significant reliance on the preparation circumstances. An additional constraining element pertains to the geometric imprecision in relation to the incline angle of the ramp. The recent achievement in producing YBCO-Au-Nb ramp-type intersections has opened up the possibility of utilizing the π phase change associated with d-wave balance as a novel useful principle for developing electronic devices. CAM junctions refer to c-axis microbridges, which are interconnections along the c-axis between a pair of superconductors that are partitioned by an insulating material.

The geometry in question exhibits an aesthetically pleasing form and possesses a low level of parasitic inductance, rendering it well-suited for implementation in multilayer circuitry. The creation of these junctions may be achieved through either the process of planarization within the mesa or through growing into an opening within an insulator. The action of the intersection is predicated upon an interface that is rich in Y, which has been created through the ion milling process of the YBCO substrate. This phenomenon is attributable to the varying milling rates that produce the cations. This categorizes the mesa intersections as interface-created junctions that share comparable constraints. Geometric unpredictability may arise in this context due to the fabrication of contacts with submicron dimensions (Putilin et al., 2000; Gao et al., 1994; Kinoshita et al., 1992).

Step-edge intersections are currently employed in circuits with one or two junctions, including SQUIDs (Sato et al., 1996; Kumakura et al., 1995). Previous attempts to effectively regulate the growth patterns of HTS films within the vicinity of the substratum step with the aim of achieving enhanced reproducibility weren't successful. The issue at hand pertains to the stability within the junctions.

SNS step-edge intersections utilizing Ag-Au alloy as the normal metal have been employed for SQUID magnetic sensors as well. The stochastic travel of the supercurrent observed in its transfer from the typical metal into the YBCO sensors as well as vice versa, which resembles a point-contact-like behavior, suggests that achieving a systematic enhancement to the repeatability of these connections may be challenging (Zoller et al., 1995; Feng et al., 2002).

Bicrystal intersections are commonly employed in the construction of demonstrator circuits. The production of these can be achieved through a single film depositing process, followed by a straightforward photolithographic step, with a satisfactory level of replication. Multilayer technology has enabled the realization of stacked junctions, as reported in references 113 and 143. Regarding commercial circuits, the expenses associated with biocrystal materials along with the design limitations resulting from the placement of contacts through the edge of the grain seem to pose a significant obstacle. The uniformity limit regarding Jc at the smallest scale is determined by the strolling of the grain borders present within the deposited high-temperature superconducting (HTS) films. This meandering seems intrinsically linked to the development of HTS film (Wagner et al., 1993). Additionally, the grain border geometry is influenced by the d-wave uniformity. The capacity for junction variables to be reproduced appears to be constrained by imperfections within the biocrystal grain borders of the substrates, resulting in an erratic trajectory characterized by meandering (Lösch et al., 1991)

The technological fascination with biepitaxial connections has waned due to the development of just 45° turn grain-boundary connections exhibiting low Jc as well as IcRN values, which have reached a relatively reproducible stage. Regrettably, the geometry under consideration exhibits pathological characteristics because of the d-wave balance inherent in the superconductive wave function of high-temperature superconductors (Yamauchi et al., 1996).

The implementation of single-layer electronic circuits has been achieved through the utilization of junctions established on weak points introduced by concentrated exposure to electron beams, which have attained a notable degree of microscopic interaction uniformity. The contacts have the capability to be situated at any position within the chip. When used in conjunction with focused ion-beam etching, the contacts enable both absolute as well as relative placement with a precision of approximately 50 nm and 1 nm, respectively. The precise stabilizing in temperature is necessary due to the significant temperature dependence exhibited by the Ic numbers. The electron beam's writing speed being slow (~ 1 μm/min), necessitates a "manufacturing time" that is roughly 1 minute for each contact. Regarding an analogous methodology relying on ion placement (Tamura et al., 1997), it remains uncertain whether the production of junctions is compatible with multiple-layer processing due to the potential impact of treatment on the underlying layering of YBCO along with the ensuing influence of high-temperature processes on previously constructed junctions through annealing.

The utilization of microbridges to traverse substrate areas that have been impaired by a concentrated ion beam (FIB), specifically the Ga-ion beam, has been observed to exhibit Josephson contact-like characteristics. This behavior is attributed to the anomalous growth patterns of YBCO films upon the substrate regions, which serve as the foundation for the microbridges (Iyo et al., 2001). According to reference 163, it is uncommon to find junctions that exhibit an ideal IV characteristics like that of an RSJ.

Traditional SIS sandwich connections may still hold promise, provided that atomic-scale management of oxide deposition occurs (Iyo et al., 2001).

In the context of radio frequency (rf) oscillators alongside other influential devices, inherent junctions created by Josephson within mesa structures of bismuth-based high-temperature superconductors (Bi-HTS) or thallium-based high-temperature superconductors (Tl-HTS) are a subject of interest, particularly in relation to non-YBCO connections. The primary issue at hand pertains to the microwave linking structure that can effectively overcome the improper impedance match found in traditional microwave circuits, rather than the setup of the intersections which were previously done using the thin film method (Jover et al., 1996; Chen et al., 1993).

4.7. WIRE AND TAPE CONDUCTORS

The Bi-HTS/Ag adhesives were the initial feasible high-temperature superconducting conductors. These tapes have since advanced to a commercial level. Tape specimens possessing cross-sectional regions measuring approximately 1 square millimeter have been developed, which can effectively convey essential currents Ic (77°K, 0 Tesla) of up to 130 amperes over distances of several kilometers. The "Powder-in-Tube" manufacturing process involves the utilization of Ag tubes which are packed with Bi-HTS precursors powder. These tubes undergo a series of thermomechanical manufacturing processes, which result in the production of tapes that contain the Bi-HTS substance in the form of thin filaments after the tubes have been flattened out. The physical basis for the effective mechanical line of Bi-HTS grains through rolling or just pressing, which results in the preferred orientation of the grains via ab-planes parallel towards the tape surface, determines the fragility regarding the Bi-22(n-1) n grains together adjacent Bi-oxide planes. The metal silver has become the sole matrix component that does not undergo chemical reactions with HTS substances and facilitates adequate oxygen dissemination. The process of Ag and Bi-HTS melting at approximately 850°C results in the formation of a strong mechanical along with electrical connection, which facilitates the bridging of non-superconducting areas through low-resistance shorts. The cost and malleability of silver are significant factors that must be considered in technical uses.

The two types of Bi-HTS systems, namely Bi-2212 and Bi-2223, are considered feasible. Bi-2212 possesses a benefit over Bi-2223 in terms of better control over chemical phase expansion. The formation of Bi-2212 is achieved through the utilization of a precursor that is of a singular phase, whereas the production of Bi-2223 necessitates the use of a starting mixture that comprises multiple phases. Currently, it is possible to attain a relatively uniform superconductive linkage among the HTS grains within Bi-2212 filaments (Antipov et al., 2002). This results in engineering essential current densities (determined in relation to the entire conductor traverse portion) Jeng (4.2 K, 0 T) > 70 kA/cm^2 across conductor lengths of several hundred meters [180]. A feasible wire conductor was discovered to achieve a current density greater than 20 kA/cm^2, regardless of the axis of the magnetic field that is generated, through the skillful configuration of Bi-2212 tapes by Jeng at a temperature of 4.2 K and magnetic field strength of 24 T. Superconducting essential current density levels Jc (4.2 K, 10 T) greater than 500 kA/cm^2 were achieved in Bi-2212 tapes through the implementation of novel processing techniques. These values are calculated in relation to the HTS portion that comprises the cross-section conductors and is approaching the threshold of epitaxial thin films, as reported in reference 181. The aforementioned findings exhibit favorable prospects for the Bi-2212 conductors in relation to its potential utilization under strong magnetic environments under low-temperature conditions (Acha et al., 1997; Tatsuki et al., 1996).

The Bi-2223/Ag adhesives have been a focal point in the creation of high-temperature superconductor conductors due to their Tc value exceeding 100 K, which enables them to function at the melting point of liquid nitrogen. The percolative superconductive flux of current within Bi-2223 filaments appears to be a challenging obstacle due to the complex phase evolution of this material, in contrast to Bi2212. In addition, the maximum magnetic field capacity of approximately 1 Tesla at a temperature of 77°K (Stangl et al., 1996), imposes limitations on the utilization of Bi-2223 tapes under liquid nitrogen (LN2) operation, thereby confining their use to low-field implementations like cables and transformers. In order to achieve operation at electromagnetic capacities of several Tesla that are of technical interest, it is necessary to cool Bi-2223 windings to a temperature of less than 30 K. Nevertheless, within this temperature range, Bi-2212 conductors could be a viable option owing to their simpler processing, resulting in reduced fabrication expenses (Klamut et al., 2001). The economic rationale has consistently supported the utilization of high-temperature superconducting (HTS) coating for uncomplicated and durable metal carrier tapes. The implementation of YBCO lightweight film methods for the production of conductors has been a challenging task due to the necessity of achieving uninterrupted biaxial texture throughout the entire length belonging to the conductor. A current density of approximately 1 MA/cm^2 is attained over a distance of roughly 10 meters, enabling initial practical demonstrations. This occurs at a temperature of 77 K and a magnetic field of 0 T. The current state of development regarding conductor length is analogous to the state of Bi-HTS conducting materials a decade ago. The materials Tl-1223 and Hg-1212 are currently being examined for their potential use in high-temperature superconducting coatings that exhibit a critical temperature exceeding 100 K. Notwithstanding their potential, the current impracticality of the necessary thin film methodologies implies that these high-temperature superconductors (HTS) do not currently constitute viable substitutes for YBCO. Previous attempts to produce Tl-1223 conductors using processing techniques designed for Bi-HTS conductive materials have yielded unfavorable outcomes (Lorenz et al., 2001; Kawashima et al., 1995).

4.7.1. Bi-2212

It is possible to manufacture Bi-2212 tape conducting materials featuring a high-temperature superconductor (HTS) fill ratio that is as high as 60% through the uncomplicated process of applying a surface encapsulating of silver bands. Tapes exhibiting Jc (4.2 K, 10 T) > 500 kA/cm^2 were produced through the implementation of enhanced processing techniques, specifically "Pre-Annealing as well as Intermediate Rolling" (PAIR). The limitations imposed by the trade-off across Jc along with the thickness within the Bi-2212 piles have resulted in restricted expectations over Jeng (4.2 K, 20 T) regarding tailored long tapes, which are anticipated to be in the range of 50–60 kA/cm^2 (Bozovic et al., 2002).

A tape conductor has been constructed in a cable resembling the "Rutherford cable," which is the conventional conductor utilized for acceleration magnets, as a real-world example. The structure was comprised of 427 filaments of Bi2212, encased in an outer covering made of AgMgSb alloy, which was selected for its superior stress resistance in comparison to pure Ag (Yamamoto et al., 2015). The wire demonstrated exceptional characteristics, with a critical current (Ic) of 285 A and a critical current density (Jc) of 340 kA/cm^2 at a temperature of 4.2 K as well as under self-field conditions. Additionally, it displayed a tensile force of 180 MPa at ambient temperature. A complete cable was found to have a critical current (Ic) of 3500 A at 4.2 K under self-field conditions. In contrast, 17-meter-long segments taken from an uninterrupted 80-meter-long cable exhibited a critical current (Ic) greater than 2700 A at 4.2 K under self-field conditions. In addition to its potential application in the development of forthcoming accelerator magnets, the connection is being examined for its utility in Superconducting Magnetized Energy storage systems ("SMES") (Naito & Hepp, 2000).

A wire-based conductor exhibiting excessive Jeng (4.2 K, 24 T) > 20 kA/cm^2 regardless of the axis of the magnetic force being applied is being developed through the careful placement of Bi-2212 tapes. This conductor, known as "ROtation-Symmetric Set up Tape-in tubes wire" or "ROSATwire," is well-suited to produce traditional solenoid magnets. Alff et al. (1999) state that it is possible to enhance the fill factor of Bi-HTS up to approximately 40%. The objective of this endeavor is to incorporate insertion magnets into 30 T class magnets, whereby the Bi-HTS coils are required to produce an extra field of multiple Tesla within an existing field of approximately 20 T. The utilization of magnet systems to perform high-field NMR represents a highly ambitious objective. Achieving temporal stability to a high degree may only be feasible through the utilization of a superconducting magnetic field system that operates in persistent mode, relying on the gradual deterioration of supercurrents (Jin et al., 1993). Nevertheless, the deteriorating times that have been shown are still inadequate to attain this objective.

4.7.2. Bi-2223

The performance of Bi-2223 Powder-in-Tube adhesive has exhibited a consistent upward trend for a duration exceeding a decade. Short specimens were able to achieve critical current density values of Jc (77 K, 0 T) ~ 70 kA/ cm^2 within HTS filaments (Iyo et al., 2001). Enhanced comprehension of the mechanically deformed mechanism along with the evolution of the Bi-2223 stage during heat annealing facilitated the manufacture of multiple 10 km tape conducting materials with a Jeng (77 K, 0 T) of approximately 15 kA/ cm^2, as well as an overall conductor cross-sectional area of roughly 1 mm^2. The maximum potential for Bi-2223/Ag tapes appears to be as high as 180 kA/cm^2, as observed locally at Jc (77 K, 0 T). The percolative supercurrent movement through the Bi-2223 grains is attributed to non-superconducting additions along with pores present in the HTS filaments. These grains are interconnected but only constitute approximately two-thirds of the filament cross-sectional area. It is anticipated that a maximum inherent limit Jeng (77 K, 0 T) of approximately 60 kA/cm^2 must be expected over high-Jc Bi-2223 conductors that have an HTS fills ratio of no more than 35%. This limit remains similar to the calculated limit of approximately 100 kA/cm^2 for YBCO-coated conductors, provided that there is a potential improvement of Jc (77 K, 0 T) to approximately 1 MA/cm^2 within YBCO coverings that are 5 μm thick upon 50 μm thick metallic carrier tapes (Jung et al., 2001; Harshman et al., 1992).

Nevertheless, in order to utilize the material in higher magnetic fields, it is imperative to operate at significantly lower temperatures. This has the added advantage of increasing the critical current density (Jc) by up to an average of seven in comparison to 77 K. Specifically, Jc (T, 0 T) ≈ Jc (77 K, 0 T) [7 – T/13 K]. An additional economic drawback pertaining to YBCO-coated conducting materials pertains to the obligatory utilization of silver to be the matrix material. Earlier cost projections for full-scale manufacturing operations were overly optimistic at 10 $/kA m, given that current estimates range from 200 to 300 $/kA m. This has been acknowledged in subsequent research, and updated projections for 2006 now range from 25 to 50 $/kA m. In addition, the effective electrical connection between the HTS filaments and the Ag matrix facilitates the connection of separated HTS regions through low-resistance shorts (Plakida et al., 1995). However, the frequent implementation of such current rerouting hinders the persistent-mode functioning of Bi-2223 coils. The Bi-2223 conducting materials possess successfully surmounted numerous obstacles in their journey toward becoming a high-temperature superconductive material with practical applications. The issue pertaining to the suppleness of the Ag structure has been effectively addressed through the implementation of dispersed MgO within the Ag structure. Alternatively, the incorporation of a slender coating of steel alloy reinforcement on both edges of the adhesives has facilitated the tapes' ability to endure a tensile load of 300 MPa along with a bending strain of 0.45% at 77 K (Tanaka et al., 2000; Watanabe et al., 2000). Several effective methods have been devised to address the issue of ac losses that stem from the electromagnetic interconnection of the HTS strings of filaments. This matter is pertinent to all power-related applications, as it dictates the requisite cooling capacity (Tsuei et al., 2000; Tsuei & Kirtley, 2002; Hilgenkamp et al., 1996). The implementation of insulating obstacles, specifically BaZrO3 or SrCO3, surrounding the HTS filaments, along with the utilization of higher resistant Ag alloys, including AgAu or just AgPd, to be matrix material, coupled with the twirling of the filaments, with twist pitches as low as lp ~ 1 cm, has been observed to effectively decrease the occurrence of coupling currents. The implementation of a unique configuration of Bi-2223 tapes in both vertical as well as horizontal stacks results in a decreased Ic-anisotropy in relation to the alignment of an externally applied magnetic field (Nilsson et al., 1994; Götz, 2000; Hilgenkamp & Mannhart, 2002).

To summarize, Bi-2223 conductors reached a pragmatic level of technological feasibility, albeit at a relatively elevated cost. Currently, the primary psychological obstacle faced by Bi-HTS conductors pertains to the high expectations surrounding the imminent arrival of a more cost-effective and superior YBCO-coated conductor (Blatter et al., 1994; Likharev & Lukens, 1988).

4.7.3. YBCO

Metal tapes covered with YBCO exhibit promising characteristics in terms of providing a cost-effective solution for high-temperature superconductor conductors that can operate at liquid nitrogen temperature, regardless of the presence of magnetic fields of several Tesla. Several YBCO coating techniques have been utilized to attain a Jc value of 1–2 MA/cm^2 in brief samples, each employing distinct approaches for achieving biaxial in nature YBCO texture. A viable approach involves the application of an ion beam to facilitate orientation-selective development during the laying of a textured buffering layer onto an untextured metallic substrate. High-Jc tapes with a length of 10 meters have been successfully demonstrated, utilizing stainless steel to provide a desirable metal option (Betouras & Joynt, 1995). An alternative approach involves the utilization of metal bands with a cube structure. To counteract the lattice mismatch while preventing the potential degradation of superconductivity within the YBCO covering due to the infiltration of metallic atoms, it is necessary to implement buffer layers such as CeO_2, Y-stabilized ZrO_2 (YSZ), $Gd_2Zr_2O_7$ (GZO), along with In-Sn oxide (ITO). The present study employs Ni as well as Ni alloys as metallic substrates that exhibit the requisite cube texture following suitable metallurgical and thermal processing (Alarco & Olsson, 1995).

The majority of buffer content that has been tested exhibits insulating properties, resulting in the inability to establish a low-resistance brief via the metallic tape that creates an interruption within the superconductive current path, unlike the Bi-HTS/Ag tapes. Viable resolutions to this issue involve the application of a layer for Au over the YBCO covering or the utilization of buffering layers composed of metallic oxides (Agassi et al., 2002).

YBCO coatings using a thickness of approximately 3 μm have been observed to exhibit a critical current density (Jc) of 1 MA/cm^2 at a temperature of 77 K and zero teslas. The utilization of high-temperature superconductors in metallic carrier tapes with a thickness of approximately 100 μm leads to a fill factor of merely 3%. The present concentration Jeng, pertaining to engineering, has been computed in relation to the complete cross-sectional area of the conductors. The value for Jeng (77 K, 0 T) is approximately 30 kA/cm^2, which is comparable to the current density attained in contemporary Bi-2223/Ag tapes that are commercially available. Metal carrier adhesives with a thickness of 50 μm are currently obtainable. The incorporation of Ca along with the refinement of the grain structure has been suggested as potential strategies for achieving the conversion of the notable intra-grain Jc (77 K, 0 T) value of approximately 5 MA/cm^2 into current densities at the macroscopic level, even in instances where the grain orientation is as low as 10° (Köster et al., 2002).

Tapes with high critical current density (Jc) values have been produced using Ion Beam Assisted Deposition (IBAD) technique.

Specifically, Jc values of 2.2 MA/cm^2 (with Ic of 78 A at 77K and 0 T) were determined over a length of 10 m, while Jc values of 0.5 MA/cm^2 were achieved over a length of 46 m (Hammerl et al., 2000). The process of depositing a YSZ buffer with a thickness of approximately 1 μm and achieving the desired biaxial texture through the use of an extra ion beam constitutes a laborious and time-intensive task. The issue can be resolved through the implementation of MgO-IBAD buffering, whereby a buffer with a thickness of approximately 10 nm is adequate for achieving biaxial texture, as stated in reference 226. The technique was successful in achieving Jc (77K, 0 T) = 0.9 MA/cm^2 (Ic (77K, 0 T) = 14.4 A) throughout a distance of 0.8 m (Schneider et al., 1999). Currently, the YBCO laser accumulation process is identified as the rate-limiting phase in fabrication, with coating rates ranging as much as 60–70 nm × m^2/h. The economic aspect of laser accumulation is a matter of concern. According to reference 218, Bi-2223/Ag tapes are comparatively less expensive than IBAD-YBCO tapes.

The process of achieving biaxial accordance with the YSZ buffering without the aid of an extra ion beam is known as Slanting Substrate Deposition ("ISD"). This is accomplished through the use of high-rate laser accumulation and proper inclination for the metallic substrate in relation to the laser plume (Hilgenkamp et al., 1999; Tallon & Loram, 2001). The deposition rate of the YSZ buffering is accelerated by a factor of 10 in comparison to IBAD. Nevertheless, the level of biaxial alignment has not reached the required threshold. To date, the attainment of Jc values of 77 K and 0 T at a level of 1 MA/cm^2 remains limited to brief samples (Varma, 1997; Paasi et al., 2001).

The Rolling-Assisted-Biaxially-Textured-Substrates (RABiTS) method involves the generation of cube-texture within a Ni (alloy) substrate adhesive through traditional rolling, which induces heavy deformation exceeding 95%. This process results in the development of a traditional roll-textured adhesive, which is then subjected to an annealing step that facilitates recrystallization through the desired bidirectionally textured cubic stage. Advanced buffering methods have been experimented with to prevent or eliminate the oxide film on the exterior of the adhesive tape. This is because the oxide layer can compromise the distinctive biaxial introductions for the substrate as well as potentially damage the YBCO covering as a result of the increase in volume that occurs during oxidation. A buffering system consisting of CeO2/YSZ/CeO^2 has been utilized to attain a Jc (77 K, 0 T) value greater than 1 MA/cm^2 in small samples. However, over an adhesive length of 0.8 m, the Jc (77 K, 0 T) value was found to be approximately 0.6 MA/cm^2 (Usoskin et al., 2002).

The "Surface-Oxidation Epitaxy" (SOE) method involves utilizing cube-textured natural Ni tape that is subjected to regulated oxidation to produce cube-textured NiO, as reported in reference 232. The YBCO coating perceives the texture of NiO, which is appropriately lattice-matched with YBCO, rather than the texture of Ni. The current density of 0.1 MA/cm^2 achieved in brief samples with supplemental MgO protecting at a temperature of 77 K and zero Tesla's magnetic field, has yet to meet the desired level of satisfaction (Kiss et al., 2001).

The investigation of utilizing Ag-substrates with biaxial texture over YBCO coating. The incorporation of costly and delicate Ag carrier tapes using the low fill rate used in the tape wrapping method does not appear to be a favorable approach. Furthermore, the number of textured grains attained is still significantly lower than the almost 100% level that has already been achieved for Ni as well as Ni metal tapes (Prusseit et al., 2000).

The utilization of deposition methods that are less expensive compared to technically proficient, yet costly vacuum vaporization techniques can result in significant cost reduction. The latest advancements in YBCO coating techniques, such as dip covering, spray pyrolysis, and sol-gel, have shown significant potential in producing specimens with an initial Jc (77 K, 0 T) of approximately 1 MA/cm^2. It has been reported that there are recent advancements in MOD-YBCO covered tapes, which have demonstrated a critical current (Ic) of 140A over a tape length of 7.5 meters at a temperature of 77 K and a magnetic field of 0 T. Liquid Stage Epitaxy serves as a potentially viable alternative (Matijasevic et al., 1997).

4.8. BULK MATERIAL

Reproducible growth of Melt-textured YBCO pellets exhibiting favorable superconducting characteristics (Jc (77 K, 1 T) > 10 kA/cm^2, Jc (50 K, 10 T) > 100 kA/cm^2) is being demonstrated in complex forms with diameters up to 6 cm. This achievement is comparable to that of NdFeB permanent magnetic materials. Currently, the manufacturing cost, which includes quality assurance, of this material exceeds 1000 EUR per kilogram. This cost is approximately ten times higher than that of NdFeB. However, it is anticipated that this cost differential will decrease once an annual output equal to or greater than tons annually is achieved. Previous studies have successfully achieved sizes up to d = 10 cm. However, it appears that connecting smaller size particles with favorable superconducting characteristics of the links may present a more feasible approach. As the size of pellets increases, achieving full oxidation turns into progressively challenging, despite the fact that it is not reliant on oxygen mass diffusion. Rather, it is dependent on oxygen transportation through microcracks (Semerad et al., 2002).

Strong pinning phenomena facilitate the immobilization of intense magnetic fields. An inductive force of 1.3 T may be established at a temperature of 77 K within single-field cylinders having a circumference of 30 mm. Additionally, Sudoh et al. (2002) report that initiation of 16 T was attained at a temperature of 24 K among two specimens with a size of 2.2 cm. More recently, reference has reported an initiation of 17 T at a temperature of 29 K between two samples with a size of 2.65 cm and a height of 1.5 cm. The aforementioned phenomenon surpasses the capabilities of traditional permanent magnets that rely on ferromagnetic spinning by a factor of ten. It is suitable for application in high-field "cryo magnets." The current limitation of YBCO cryomagnets isn't due to pinning, but rather due to the mechanical stresses involved. These forces surpass the breaking strength of the YBCO ceramics in high domains, which is typically between 20–30 MPa over YBCO lacking any additives. This results in the cracking of pellets unless appropriate mechanical stability is implemented. The inclusion of Ag within the range of 246 to 250 has been demonstrated to confer advantageous effects in terms of enhancing intrinsic stability. The application of resin impregnation for passivation purposes is known to provide protection against moisture, thereby enhancing the durability of the samples over extended periods. The integration of melt-textured YBCO and permanent magnets results in levitation characteristics that are appealing for utilization in bearing apps. Superconducting magnetic drives exhibit self-stabilization owing to transition pinning, compared to conventional magnetic drives. The levitation prompts exhibited by YBCO pellets with a diameter of 3 cm and zero field cooling are significant, exceeding 80 N at an offset of 0.5 mm across 0.4 T/d = 2.5 cm SmCo magnets at 77 K. These forces possess force densities greater than 10 N/cm^2 and are constrained by the magnetism strength generated by the enduring magnet, as reported in references 252, 253, and 254. The rigidity of the levitation factors is additionally restricted through the magnetized field power. YBCO particles are being evaluated as a potential replacement for NbTi-coils cooled with liquid helium within the Japanese Maglev undertaking (Aidam et al., 2001).

Unlike YBCO, the production of excellent rare-earth 123 specimens necessitate a processing environment with reduced oxygen and cannot be achieved in ambient air. Enhanced superconducting characteristics have been documented over melt-textured Nd-123 as well as Sm123, albeit solely for specimens with dimensions up to approximately 3 cm (d). However, the attainment of an optimal initiation of 2.0 T within Gd123 specimens with a diameter of approximately 5 cm at a temperature of 77 K, as well as the confinement of 3.3 T within two such specimens at the same temperature, surpasses the accomplishments made with YBCO. The pinning of the more powerful magnetic fields at 77 K has been observed to be even more effective in ternary combinations of Rare-Earth-123. Commercially accessible Bi-2212 rods as well as tubes exhibit electrical properties that are suitable for various applications, including but not limited to current leads along with defective current limiters. These rods along with tubes are offered in measures up to a diameter of 30 cm along with a length of 50 cm (Kinder et al., 1997).

Activity 1.1

You are part of a research team tasked with optimizing high-temperature superconductors for a specific application, such as energy transmission. Design a simulation or modeling activity that allows you to test different material properties and their effects on the efficiency and performance of the superconducting system.

4.9. SUMMARY

Thus chapter explores the materials aspects of high-temperature superconductors (HTS) and their relevance in practical applications. It begins with an introduction to the chapter's topics and then delves into the atomic structure and classification of HTS materials. The chapter highlights the structural obstacles that can impede supercurrents, which are critical for the optimal performance of high-temperature superconductors. Furthermore, the chapter discusses technically applicable HTS materials, including thin films, Josephson junctions, wire and tape conductors (such as Bi-2212, Bi-2223, and YBCO), and bulk materials. It provides insights into the properties and characteristics of these materials and their potential applications in various technological fields.

REVIEW QUESTIONS

1. Why is understanding the materials aspect of high-temperature superconductors important for practical applications?
2. Explain the significance of structural obstacles for supercurrents in high-temperature superconductors.
3. What are some examples of technically applicable HTS materials mentioned in this chapter?
4. Discuss the advantages of thin films in high-temperature superconductivity and their potential applications.

5. What are Josephson junctions, and how are they relevant to high-temperature superconductors?
6. Describe the characteristics and properties of wire and tape conductors in high-temperature superconductivity.
7. Provide an overview of Bi-2212 and its use in wire and tape conductors.
8. How does Bi-2223 contribute to the field of high-temperature superconductivity?
9. Discuss the properties and applications of YBCO (yttrium barium copper oxide) in high-temperature superconductors.

REFERENCES

1. Acha, C., Loureiro, S. M., Chaillout, C., Tholence, J. L., Capponi, J. J., Marezio, M., & Nunez-Regueiro, M. (1997). 50 K enhancement of Tc by pressure in the Hg-2212 superconductor. *Solid State Communications, 102*(1), 1–5.
2. Agassi, D., Christen, D. K., & Pennycook, S. J. (2002). Flux pinning and critical currents at low-angle grain boundaries in high-temperature superconductors. *Applied Physics Letters, 81*(15), 2803–2805.
3. Ahn, C. H., Triscone, J. M., Archibald, N., Decroux, M., Hammond, R. H., Geballe, T. H., & Beasley, M. R. (1995). Ferroelectric field effect in epitaxial thin film oxide $SrCuO_2$/Pb (Zr_0. 52Ti0. 48) O_3 heterostructures. *Science, 269*(5222), 373–376.
4. Aidam, R., Geerk, J., Linker, G., Ratzel, F., Reiner, J., Schneider, R., & Schwab, R. (2001). Y-Ba-Cu-O thin films on 3" sapphire wafers for microwave devices. *IEEE Transactions on Applied Superconductivity, 11*(1), 357–360.
5. Ai-Mamouri, M., Edwards, P. P., Greaves, C., & Slaski, M. (1994). Synthesis and superconducting properties of the strontium copper oxy-fluoride $Sr_2CuO_2F^{2+}$ δ. *Nature, 369*(6479), 382–384.
6. Alarco, J. A., & Olsson, E. (1995). Analysis and prediction of the critical current density across [001]-tilt $YBa_2Cu_3O_{7-\delta}$ grain boundaries of arbitrary misorientation angles. *Physical Review B, 52*(18), 13625.
7. Anderson, P. W. (1987). The resonating valence bond state in La_2CuO_4 and superconductivity. *Science, 235*(4793), 1196–1198.
8. Alff, L., Meyer, S., Kleefisch, S., Schoop, U., Marx, A., Sato, H., & Gross, R. (1999). Anomalous Low Temperature Behavior of Superconducting $Nd_{1.85}Ce_{0.15}CuO_{4-y}$. *Physical Review Letters, 83*(13), 2644
9. Anderson, P. W. (1997). *The Theory of Superconductivity in the High-Tc Cuprate Superconductors, 1(1), 4-9.*
10. Antipov, E. V., Abakumov, A. M., & Putilin, S. N. (2002). Chemistry and structure of Hg-based superconducting Cu mixed oxides. *Superconductor Science and Technology, 15*(7), R31.
11. Batlogg, B. (1998). Cuprate superconductors: science beyond high Tc. *Solid State Communications, 107*(11), 639–647.
12. Batlogg, B., Hwang, H. Y., Takagi, H., Cava, R. J., Kao, H. L., & Kwo, J. (1994). Normal state phase diagram of (La, $Sr)_2CuO_4$ from charge and spin dynamics. *Physica C: Superconductivity, 235,* 130–133.
13. Bednorz, J. G., & Müller, K. A. (1986). Possible high T c superconductivity

in the Ba–La–Cu–O system. *Zeitschrift für Physik B Condensed Matter, 64*(2), 189–193.

14. Betouras, J., & Joynt, R. (1995). Theoretical study of the critical current of $YBa_2Cu_3O_{7-\delta}$ bicrystals with hole-deficient grain boundaries. *Physica C: Superconductivity, 250*(3–4), 256–264
15. Blatter, G., Feigel'man, M. V., Geshkenbein, V. B., Larkin, A. I., & Vinokur, V. M. (1994). Vortices in high-temperature superconductors. *Reviews of Modern Physics, 66*(4), 1125.
16. Bordet, P., LeFloch, S., Chaillout, C., Duc, F., Gorius, M. F., Perroux, M., & Tholence, J. L. (1997). AuBa2 (Y_{1-x}, Ca_x) Cu_2O_7: A new superconducting gold cuprate with Tc above 80 K. *Physica C: Superconductivity, 276*(3–4), 237–244.
17. Bormann, R. & Nölting, J. (1989). Stability limits of the perovskite structure in the Y-Ba-Cu-O system. *Applied physics letters, 54*(21), 2148-2150.
18. Bozovic, I., Logvenov, G., Belca, I., Narimbetov, B., & Sveklo, I. (2002). Epitaxial Strain and Superconductivity in $La_{2-x}Sr_xCuO_4$ Thin Films. *Physical Review Letters, 89*(10), 107001
19. Brosha, E. L., Davies, P. K., Garzon, F. H., & Raistrick, I. D. (1993). Metastability of superconducting compounds in the Y-Ba-Cu-O system. *Science, 260*(5105), 196–198.
20. Chen, Z. Y., Sheng, Z. Z., Tang, Y. Q., Li, Y. F., Wang, L. M., & Pederson, D. O. (1993). Synthesis and characterization of 2234-phase $Ti_2Ba_2Ca_3Cu_4O_{12}$ prepared near and at the melting point. *Superconductor Science and Technology, 6*(4), 261.
21. Chu, C. W. (1997). High-temperature superconducting materials: a decade of impressive advancement of T/sub c. *IEEE Transactions on Applied Superconductivity, 7*(2), 80–89.
22. Chu, C. W. (1999). High-temperature superconductivity: A macroscopic view. *Journal of Superconductivity, 12*, 85–88.
23. Chu, C. W., Salama, K., Hor, P. H., & Chu, W. K. (Eds.). (1992). *HTS Materials, Bulk Processing and Bulk Applications-Proceedings of the 1992 TCSUH Workshop*. World Scientific.
24. Erb, A., Walker, E., Genoud, J. Y., & Flükiger, R. (1997). Improvements in crystal growth and crystal homogeneity and its impact on physics. *Physica C: Superconductivity, 282*, 89–92.
25. Feng, D. L., Damascelli, A., Shen, K. M., Motoyama, N., Lu, D. H., Eisaki, H., & Shen, Z. X. (2002). Electronic Structure of the Trilayer

Cuprate Superconductor $Bi_2Sr_2Ca_2Cu_3O^{10+}$ δ. *Physical Review Letters, 88*(10), 107001.

26. Gao, L., Xue, Y. Y., Chen, F., Xiong, Q., Meng, R. L., Ramirez, D., & Mao, H. K. (1994). Superconductivity up to 164 K in $HgBa_2Ca_{m-1}Cu_mO_{2m+2+\delta}$ (m = 1, 2, and 3) under quasihydrostatic pressures. *Physical Review B, 50*(6), 4260.
27. Götz, B. (2000). *Untersuchung der Transporteigenschaften von Korngrenzen in Hochtemperatur-Supraleitern, 1(2), 8-10.*
28. Hammerl, G., Schmehl, A., Schulz, R. R., Götz, B., Bielefeldt, H., Schneider, C. W., & Mannhart, J. (2000). Enhanced supercurrent density in polycrystalline $YBa_2Cu_3O_7$-δ at 77 K from calcium doping of grain boundaries. *Nature, 407*(6801), 162–164.
29. Harshman, D. R., & Mills Jr, A. P. (1992). Concerning the nature of high-T c superconductivity: Survey of experimental properties and implications for interlayer coupling. *Physical Review B, 45*(18), 10684.
30. Hilgenkamp, H., & Mannhart, J. (2002). Grain boundaries in high-T c superconductors. *Reviews of Modern Physics, 74*(2), 485.
31. Hilgenkamp, H., Mannhart, J., & Mayer, B. (1996). Implications of d x_2– y_2 symmetry and faceting for the transport properties of grain boundaries in high-T c superconductors. *Physical Review B, 53*(21), 14586
32. Hilgenkamp, H., Schneider, C. W., Schulz, R. R., Götz, B., Schmehl, A., Bielefeldt, H., & Mannhart, J. (1999). Modifying electronic properties of interfaces in high-Tc superconductors by doping. *Physica C: Superconductivity, 326*, 7–11.
33. Hiroi, Z., Kobayashi, N., & Takano, M. (1994). Probable hole-doped superconductivity without apical oxygens in (Ca, Na) 2CuO2CI2. Nature, 371(6493), 139–141.
34. Ihara, H. (2001). How to achieve the best performance superconductor based on Cu-1234. *Physica C,* 364–365.
35. Iyo, A., Aizawa, Y., Tanaka, Y., Tokumoto, M., Tokiwa, K., Watanabe, T., & Ihara, H. (2001). High-pressure synthesis of $TlBa_2$Can-1CunOy (n = 3 and 4) with Tc= 133.5 K (n = 3) and 127 K (n = 4). *Physica C: Superconductivity, 357*, 324–328
36. Iyo, A., Tanaka, Y., Ishiura, Y., Tokumoto, M., Tokiwa, K., Watanabe, T., & Ihara, H. (2001). Study on enhancement of Tc (□ 130 K) in $TlBa_2Ca_2Cu_3Oy$ superconductors. *Superconductor Science and Technology, 14*(7), 504.
37. Iyo, A., Tanaka, Y., Tokumoto, M., & Ihara, H. (2001). High-pressure synthesis and properties of Ba2Can-1CunO2n (O, F) 2 (n= 2–5)

superconductors. *Physica C: Superconductivity, 366*(1), 43–50

38. Jin, C. Q., Yao, Y. S., Liu, S. C., Zhou, W. L., & Wang, W. K. (1993). Superconductivity and microstructure of n-type Ln1. 85Ce0. 15CuO_{4-y} (Ln= Pr, Sm, Eu) produced under high-pressure sintering. *Applied Physics Letters, 62*(23), 3037–3039.
39. Johnson, J. E., Tang, C. L., & Grande, W. J. (1993). Optical flip-flop based on two-mode intensity bistability in a cross-coupled bistable laser diode. *Applied Physics Letters, 63*(24), 3273–3275.
40. Jover, D. T., Wijngaarden, R. J., Griessen, R., Haines, E. M., Tallon, J. L., & Liu, R. S. (1996). Pressure dependence of the superconducting critical temperature of $Tl_2Ba_2Ca_2Cu_3O^{10+}$ y and $Tl_2Ba_2Ca_3Cu_4O^{12+}$ y up to 21 GPa. *Physical Review B, 54*(14), 10175.
41. Jung, C. U., Kim, J. Y., Lee, S. M., Kim, M. S., Yao, Y., Lee, S. Y., & Ha, D. H. (2001). High-pressure synthesis of the homogeneous infinite-layer superconductor Sr0. 9La0. 1CuO2. *Physica C: Superconductivity and Its Applications, 364*, 225–227.
42. Kawashima, T., Matsui, Y., & Takayama-Muromachi, E. (1995). New series of oxide superconductors, $BSr_2Ca_{n-1}Cu_nO_{2n+3}$ (n= 3~ 5), prepared at high pressure. *Physica C: Superconductivity, 254*(1–2), 131–136.
43. Kawashima, T., Matsui, Y., & Takayama-Muromachi, E. (1996). New oxyfluoride superconductors $Sr_2Ca_{n-1}Cu_nO_{2n+\delta F2\pm y}$ (n= 2; Tc= 99 K, n= 3; Tc= 111 K) prepared at high pressure. *Physica C: Superconductivity and Its Applications*, *257*(3–4), 313–320.
44. Kinder, H., Berberich, P., Prusseit, W., Rieder-Zecha, S., Semerad, R., & Utz, B. (1997). YBCO film deposition on very large areas up to 20× 20 cm2. *Physica C: Superconductivity, 282*, 107–110.
45. Kinoshita, K., & Yamada, T. (1992). A new copper oxide superconductor containing carbon. *Nature, 357*(6376), 313–315.
46. Kiss, T., Noda, S., Nishimura, S., Ohya, K., & Utsunomiya, D. (2001). Yu. A. llyn, H. Okamoto. *Physica C, 357*(360), 1165.
47. Klamut, P. W., Dabrowski, B., Mini, S. M., Maxwell, M., Kolesnik, S., Mais, J., & Xiao, Y. (2001). Magnetic properties of RuSr2RECu2O8 (RE= Gd, Eu) and $Ru_{1-x}Sr_2GdCu_{2+x}O_{8-y}$ superconductors. *Physica C: Superconductivity and Its Applications, 364*, 313–319.
48. Köster, S. A., Moshnyaga, V., Samwer, K., Lebedev, O. I., Tendeloo, G. V., Shapoval, O., & Belenchuk, A. (2002). Doping of interfaces in (La 0.7 Sr 0.3 $MnO_3)_{1-x}$:$(MgO)_x$ composite films. *Applied Physics Letters, 81*(9), 1648–1650.

49. Kumakura, H., Togano, K., Kawashima, T., Takayama-Muromachi, E., Okayasu, S., & Kazumata, Y. (1995). Flux pinning and irreversibility lines of new oxycarbonate superconductors (Cu/sub 0.5/C/sub 0.5/) Ba/sub 2/Ca/sub n-1/Cu/sub n/O/sub 2n+ 3/(n= 3, 4) and (Cu/sub 0.5/C/sub 0.5/)/sub 2/Ba/sub 3/Ca/sub n-1/Cu/sub n/O/sub 2n+ 3/(n= 3, 4). *IEEE Transactions on Applied Superconductivity*, *5*(2), 1399–1404.
50. Liitgemeier, H., Schmenn, S., & Meuffels, P. (1996). 0. Storz, R. Schollhorn, Ch. Niedermayer, I. Heinmaa, and Yu Baikov. *Physica C267,* 191.
51. Likharev, K. K., & Lukens, J. (1988). Dynamics of Josephson junctions and circuits. *Physics Today, 41*(11), 122.
52. Lin, J. G., Huang, C. Y., Xue, Y. Y., Chu, C. W., Cao, X. W., & Ho, J. C. (1995). Origin of the R-ion effect on Tc in RBa2Cu3O7. *Physical Review B, 51*(18), 12900.
53. Lorenz, B., Meng, R. L., Cmaidalka, J., Wang, Y. S., Lenzi, J., Xue, Y. Y., & Chu, C. W. (2001). Synthesis, characterization and physical properties of the superconducting ferromagnet RuSr2GdCu2O8. *Physica C: Superconductivity, 363*(4), 251–259.
54. Lösch, S., Budin, H., Eibl, O., Hartmann, M., Rentschler, T., Rygula, M., & Huebener, R. P. (1991). Structural analysis and superconducting properties of Pb-free Bi2Sr2Ca3Cu4Oy (n= 4, nominal composition). *Physica C: Superconductivity, 177*(4–6), 271–280.
55. Matijasevic, V., Lu, Z., Kaplan, T., Huang, C., & Zhang, D. (1997, June). Atomic-absorption controlled reactive evaporation of HTS wafers for microwave applications. In *Conference Series-Institute of Physics* (Vol. 158, pp. 189–192). IOP Publishing LTD.
56. Mawdsley, A., Tallon, J. L., & Presland, M. R. (1992). Thermodynamic stability of cuprate perovskite superconductors. *Physica C: Superconductivity, 190*(4), 437–443.
57. Moreo, A., Yunoki, S., & Dagotto, E. (1999). Phase separation scenario for manganese oxides and related materials. *Science*, *283*(5410), 2034–2040.
58. Naito, M., & Hepp, M. (2000). Superconducting T′-La2-xCexCuO4 films grown by molecular beam epitaxy. *Japanese Journal of Applied Physics, 39*(6A), L485.
59. Nilsson, P. Å., Ivanov, Z. G., Olsson, H. K., Winkler, D., Claeson, T., Stepantsov, E. A., & Tzalenchuk, A. Y. (1994). Bicrystal junctions and superconducting quantum interference devices in YBa2Cu3O7 thin films. *Journal of Applied Physics, 75*(12), 7972–7977.
60. Noe, M., Juengst, K. P., Werfel, F. N., Elschner, S., Bock, J., Wolf, A., &

Breuer, F. (2002). Measurements and tests of HTS bulk material in resistive fault current limiters. *Physica C: Superconductivity,* 372, 1626–1630.

61. Orenstein, J., & Millis, A. J. (2000). Advances in the physics of high-temperature superconductivity. *Science, 288*(5465), 468–474.
62. Paasi, J., Herrmann, P. F., Verhaege, T., Lehtonen, J., Bock, J., Cowey, L., & Collet, M. (2001). Superconducting power link for power transmission and fault current limitation. *Physica C: Superconductivity, 354*(1–4), 1–4.
63. Plakida, N. M., & Plakida, N. M. (1995). Electronic Properties of High-Tc Superconductors. *High-Temperature Superconductivity: Experiment and Theory*, 84–144.
64. Prusseit, W., Furtner, S., & Nemetschek, R. (2000). Series production of large area YBa2Cu3O7 films for microwave and electrical power applications. *Superconductor Science and Technology*, *13*(5), 519.
65. Putilin, S. N., Antipov, E. V., Abakumov, A. M., Rozova, M. G., Lokshin, K. A., Pavlov, D. A., & Marezio, M. (2000). Effect of fluorination and high pressure on the structure and properties of the Hg-bearing superconducting Cu mixed oxides. *Physica C: Superconductivity, 338*(1–2), 52–59.
66. Renner, C., Aeppli, G., Kim, B. G., Soh, Y. A., & Cheong, S. W. (2002). Atomic-scale images of charge ordering in a mixed-valence manganite. Nature, 416(6880), 518–521.
67. Salamon, M. B., & Jaime, M. (2001). The physics of manganites: Structure and transport. *Reviews of Modern Physics, 73*(3), 583.
68. Sato, M., Akimitsu, J., Takahashi, H., & Mori, N. (1996). Effect of high pressure on Tc of a homologous series of oxycarbonate superconductors Sr_2 (Ca, Sr)$_{n-1}$$Cu_n$$(CO_3)_{1-x}$ (BO_3) xO_y (n= 1, 2) and Ba2Ca2Cu3 (CO3)$_{1-x}$ (BO3) xOy. *Physica C: Superconductivity, 271*(1–2), 79–82.
69. Schneider, C. W., Schulz, R. R., Götz, B., Schmehl, A., Bielefeldt, H., Hilgenkamp, H., & Mannhart, J. (1999). Tailoring of high-T c Josephson junctions by doping their electrodes. *Applied Physics Letters, 75*(6), 850–852.
70. Semerad, R., Knauf, J., Irgmaier, K., & Prusseit, W. (2002). RE-123 thin films for microwave applications. *Physica C: Superconductivity, 378*, 1414–1418.
71. Service, R. F. (1996). Superconductivity Turns 10: A decade after their discovery, compounds that superconduct electricity at high temperatures (HTS) are making it into real-world devices-but still, no one can prove how they work. Science, *271*(5257), 1804–1806.
72. Stangl, E., Proyer, S., Borz, M., Hellebrand, B., & Bäuerle, D. (1996).

Pulsed-laser deposited GdBaSrCu3O$_{7-\delta}$ and GdBa2Cu3O$_{7-\delta}$ films. *Physica C: Superconductivity, 256*(3–4), 245–252.

73. Sudoh, K., Ichino, Y., Itoh, M., Yoshida, Y., Takai, Y., & Hirabayashi, I. (2002). Correlation between Film Thickness and Critical Current Density of Sm1+ xBa2-xCu3O6+ δ Films Deposited by Pulsed Laser Deposition. *Japanese Journal of Applied physics, 41*(9A), L983.
74. Sun, J. Z. (2001). Spin-dependent transport in trilayer junctions of doped manganites. *Physica C: Superconductivity, 350*(3–4), 215–226.
75. Tallon, J. L., & Loram, J. W. (2001). The doping dependence of T* – what is the real high-Tc phase diagram? *Physica C: Superconductivity, 349*(1–2), 53–68.
76. Tallon, J. L., Williams, G. V. M., & Loram, J. W. (2000). Factors affecting the optimal design of high-Tc superconductors—the pseudogap and critical doping. *Physica C: Superconductivity, 338*(1–2), 9–17.
77. Tamura, T., Adachi, S., Wu, X. J., Tatsuki, T., & Tanabe, K. (1997). Pb-1223 cuprate superconductor with Tc above 120 K synthesized under high pressure. *Physica C: Superconductivity, 277*(1–2), 1–6.
78. Tanaka, K., Iyo, A., Tanaka, Y., Tokiwa, K., Tokumoto, M., Ariyama, M., & Ihara, H. (2000). Low superconducting anisotropy (γ= 5–11) in (Cu, Tl)-1223 superconductors. *Physica B: Condensed Matter, 284*, 1081–1082.
79. Tatsuki, T., Tokiwa-Yamamoto, A., Fukuoka, A., Tamura, T., Wu, X. J., Moriwaki, Y., & Tanaka, S. T. S. (1996). A New Homologous Series of Superconducting Cuprates with the 22 (n-1) n Structure,(Hg1-xTlx) 2Ba2Can-1CunOy (x ~ 0.4, n = 3, 4). *Japanese Journal of Applied Physics, 35*(2B), L205.
80. Tsuei, C. C., & Kirtley, J. R. (2000). Pairing symmetry in cuprate superconductors. *Reviews of Modern Physics, 72*(4), 969.
81. Tsuei, C. C., & Kirtley, J. R. (2002). d-Wave pairing symmetry in cuprate superconductors—fundamental implications and potential applications. *Physica C: Superconductivity, 367*(1–4), 1–8.
82. Usoskin, A., Issaev, A., Freyhardt, H. C., Leghissa, M., Oomen, M. P., & Neumueller, H. W. (2002). Interchange of current between shunt layer and HTS film in long coated tapes. *Physica C: Superconductivity, 372*, 857–862.
83. Varma, C. M. (1997). Non-Fermi-liquid states and pairing instability of a general model of copper oxide metals. *Physical Review B, 55*(21), 14554.
84. Wagner, P., Hillmer, F., Frey, U., Adrian, H., Steinborn, T., Ranno, L., & Bruynseraede, Y. (1993). Preparation and structural characterization

of thin epitaxial Bi2Sr2CaCu2O8+ δ films with Tc in the 90 K range. *Physica C: Superconductivity, 215*(1–2), 123–131.

85. Watanabe, T., Miyashita, S., Ichioka, N., Tokiwa, K., Tanaka, K., Iyo, A., & Ihara, H. (2000). Carrier doping and superconducting properties in Cu-1234 and CuTl-1223 superconductors. *Physica B: Condensed Matter, 284,* 1075–1076.
86. Williams, G. V. M., & Tallon, J. L. (1996). Ion size effects on 7 and interplanar coupling in $RBa2Cu3O_{7-\delta}$. *Physica C: Superconductivity, 258*(1–2), 41–46.
87. Wu, M. K., Ashburn, J. R., Torng, C., Hor, P. H., Meng, R. L., Gao, L., & Chu, A. (1987). Superconductivity at 93 K in a new mixed-phase Y-Ba-Cu-O compound system at ambient pressure. *Physical Review Letters, 58*(9), 908.
88. Wu, N. L., Du, Z. L., Xue, Y. Y., Rusakova, I., Ross, D. K., Gao, L., & Raveau, B. (1999). Observation and identification of a 107 K superconducting family in the $CaBa2Ca2Cu3O_{9-\delta}$ system. *Physica C: Superconductivity, 315*(3–4), 227–234
89. Yamamoto, A., Hu, W. Z., Izumi, F., & Tajima, S. (2001). Superconducting and structural properties of nearly carbonate-free HgBa2CuO4+ δ. *Physica C: Superconductivity, 351*(4), 329–340.
90. Yamamoto, H., Krockenberger, Y., & Naito, M. (2015). Epitaxial growth of superconducting oxides. In *Epitaxial Growth of Complex Metal Oxides* (pp. 95–127). Woodhead Publishing.
91. Yamauchi, H., & Karppinen, M. (2000). Application of high-pressure techniques: stabilization and oxidation-state control of novel superconductive and related multi-layered copper oxides. *Superconductor Science and Technology, 13*(4), R33.
92. Yamauchi, H., Karppinen, M., & Tanaka, S. (1996). Homologous series of layered cuprates. *Physica C: Superconductivity, 263*(1–4), 146–150.
93. Yang, G., Abell, J. S., & Gough, C. E. (1999). High temperature c-axis resistivity of Bi2Sr2CaCu2O 8+ x single crystals and dependence on oxygen stoichiometry. *Applied Physics Letters, 75*(13), 1955–1957.
94. Zoller, P., Glaser, J., Ehmann, A., Schulz, C., Wischert, W., Kemmler-Sack, S., & Huebener, R. P. (1995). Superconductivity and processing of single Bi–O layered cuprates in the (Bi, Pb)–Sr–(Ca, Y)–Cu–O system. *Zeitschrift für Physik B Condensed Matter, 96*(4), 505–509

CHAPTER

5 Synthesis and Fabrication Techniques of High-Temperature Superconductors

Contents

5.1. INTRODUCTION

The techniques utilized in synthesizing and fabricating high-temperature superconductors (high-Tc) are crucial in the progress and evolution of this area. The significance of these techniques lies in their direct influence on the material's structure, composition, purity, and superconducting properties. The attainment of high-Tc superconductivity necessitates meticulous synthesis and fabrication procedures management, given that even minor deviations can substantially impact the critical temperature and other pivotal factors. The complexity of high-Tc superconductors is a key factor contributing to the importance of synthesis and fabrication techniques. In contrast to traditional superconductors, high-temperature superconducting materials display intricate crystal structures, intricate electronic interactions, and non-traditional pairing mechanisms. Therefore, synthesizing these materials with specific properties necessitates advanced techniques to accurately regulate variables such as composition, crystal structure, and defect density (Behrman, 1987).

In addition, high-temperature superconductors frequently display anisotropic characteristics, whereby their superconductivity manifests differently along distinct crystallographic axes. The degree of anisotropy can be subject to alteration based on the techniques utilized in the synthesis and fabrication processes. It becomes feasible to manipulate the crystal orientation, epitaxial quality, and interface characteristics of the material through precise adjustments to the growth or deposition conditions. This optimization of superconducting properties and performance enhancement in particular applications is achieved. The requirement for materials with high levels of purity is a critical factor in the context of high-temperature superconductors with high-Tc. The formation of Cooper pairs, which are responsible for superconductivity, can be hindered by even minor concentrations of impurities that disrupt the delicate balance of electron-electron interactions. Synthesis techniques that guarantee the purity of the initial materials and reduce the incorporation of impurities are imperative in attaining high-Tc superconductivity and optimizing critical temperatures (Logyenov et al., 2009) (Figure 5.1).

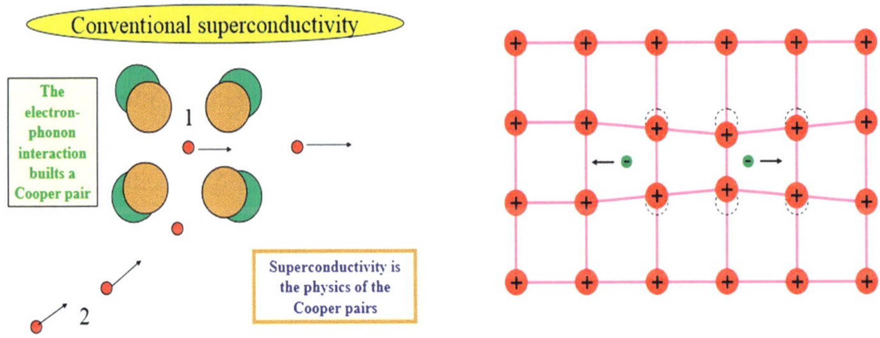

Figure 5.1. *Schematic of cooper pair formation.*

Source: Shatha Hashim Mahdi, Creative Commons License.

The practical applications of high-Tc superconductors necessitate the cruciality of scalability and reproducibility of synthesis and fabrication techniques. The potential of these materials for diverse technological applications, including energy transmission, power devices, and quantum computing, underscores the importance of devising synthesis techniques that can be readily upscaled to yield substantial quantities of superconducting materials of superior quality. The attainment of consistency and reproducibility is of utmost importance in facilitating the assimilation of high-Tc superconductors into practical applications and technological advancements (Ikram et al., 2021).

Tip

Understanding the various single crystal growth techniques is essential for producing high-quality single crystals of high-temperature superconductors with desirable properties.

The chapter on "Synthesis and Fabrication Techniques of High-Temperature Superconductors" comprehensively examines the diverse techniques and procedures utilized in producing high-Tc superconducting substances. The present chapter is a comprehensive manual encompassing conventional and state-of-the-art methodologies for synthesizing and fabricating these noteworthy materials. The chapter underscores the significance of utilizing these techniques to attain specific material properties and facilitate feasible applications of high-Tc superconductors. The introductory section of the chapter emphasizes the importance of synthesis and fabrication methodologies within the realm of high-temperature superconductors. The introductory section establishes the context for the following segments, which provide detailed insights into particular techniques and methodologies. The information is presented methodically, providing a coherent and cohesive structure (Sengupta, 1998).

The primary section of interest pertains to techniques employed for synthesizing powders. The article comprehensively analyzes several methodologies, including solid-state reaction, ceramic precursor techniques, chemical vapor deposition (CVD), sol-gel approach, and spray pyrolysis. Each methodology is illustrated with its fundamental principles, procedural steps, and its benefits and drawbacks (Figure 5.2).

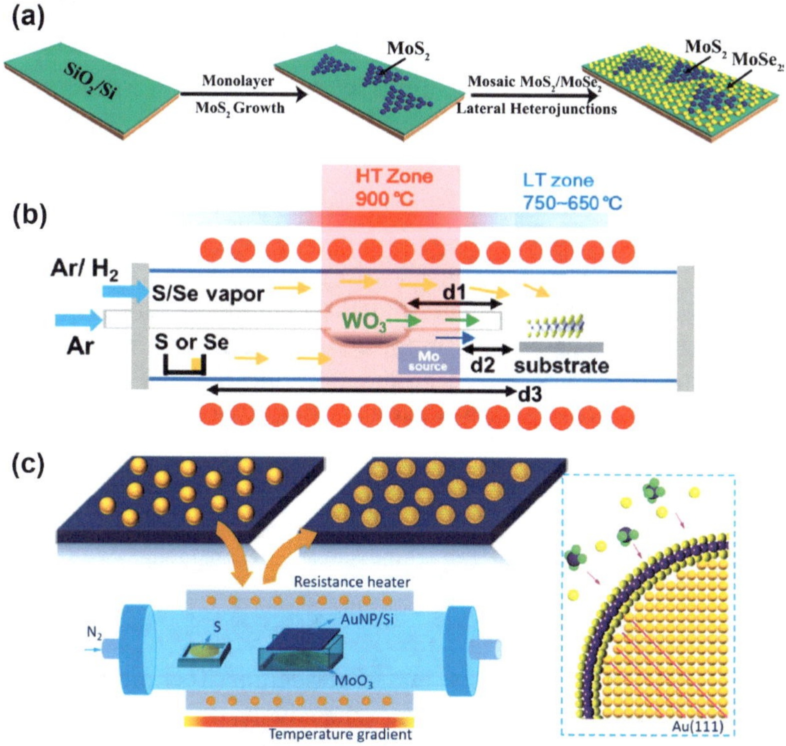

Figure 5.2. *Chemical vapor deposition (CVD) method: (a) Schematic illustration of the two-step CVD technique for the preparation of mosaic MoS_2–$MoSe_2$ lateral heterojunctions film on SiO_2/Si substrate. (b) Schematic illustration of the synthesis setup and various modes of hetero-epitaxial growth of TMD atomic layers. (c) Schematic showing the CVD process of MoS_2 shell growth on Au nanoparticles.*

Source: MKS Instruments, Creative Commons License.

The following section explores the techniques employed in the growth of single crystals, which are of utmost importance in investigating the fundamental characteristics of high-temperature superconductors. This comprehensive coverage pertains to various methods such as the flux, Bridgman-Stockbarger, traveling solvent floating zone, Czochralski, and molecular beam epitaxy (MBE). The present chapter furnishes valuable insights into crystal growth mechanisms and parameters and the challenges of pursuing high-quality single crystals (Zahi, 2010).

The subsequent section scrutinizes techniques utilized for the deposition of thin films. The article presents a comprehensive overview of various physical vapor deposition (PVD) techniques, such as thermal evaporation, sputtering, and molecular beam epitaxy (MBE), in addition to chemical vapor deposition (CVD) and atomic layer deposition (ALD) methods (Figure 5.3).

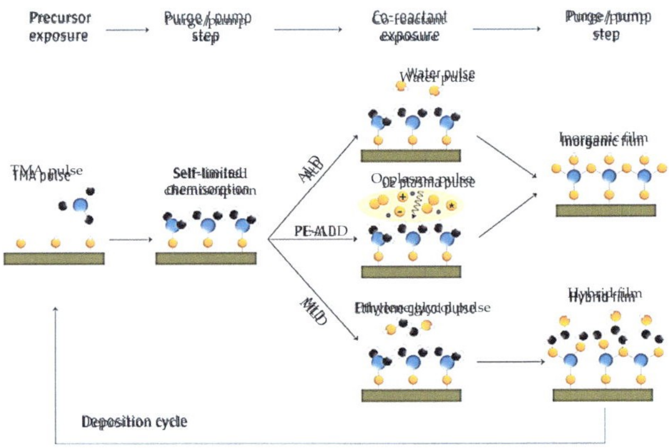

Figure 5.3. *Illustration of atomic layer deposition (ALD) methods.*

Source: O. Trejo, Creative Commons License.

The chapter delves into the complexities associated with these techniques, including the mechanisms of film formation, the selection of substrates, and the regulation of thickness and composition. The chapter examines in situ synthesis and growth methodologies, which facilitate the direct production of superconducting thin films. This study provides a comprehensive analysis of Pulsed Laser Deposition (PLD), Metal-Organic Chemical Vapor Deposition (MOCVD), Hybrid Physical-Chemical Vapor Deposition (HPCVD), and Hybrid Molecular Beam Epitaxy (MBE), with a focus on their distinctive benefits and limitations. The subsequent section deals with processing and fabrication techniques. This study investigates mechanical milling and mixing techniques, tape casting, laser ablation, and inkjet printing to shape and manipulate high-Tc superconducting materials. The chapter offers valuable insights into the intricate processes involved in material processing, the impact of processing parameters on material properties, and the appropriateness of each technique for diverse applications (Rupich et al., 2004).

A distinct section is dedicated to discussing post-processing and heat treatment techniques deemed essential for optimizing the properties of high-Tc superconductors. This study investigates the impact of various techniques on material properties, including annealing, oxygenation processes, stoichiometry optimization, quenching, rapid thermal processing, and hydrothermal treatments. The challenges associated with each method are also analyzed. The chapter additionally discusses the characterization of materials that have been synthesized. The present study investigates various methodologies, namely X-ray diffraction (XRD), scanning electron microscopy (SEM), transmission electron microscopy (TEM), energy-dispersive X-ray spectroscopy (EDX), and Raman spectroscopy. The chapter emphasizes the significance of conducting a thorough characterization to verify high-temperature superconducting materials' intended structural, morphological, and chemical attributes (Namburi et al., 2021) (Figure 5.4).

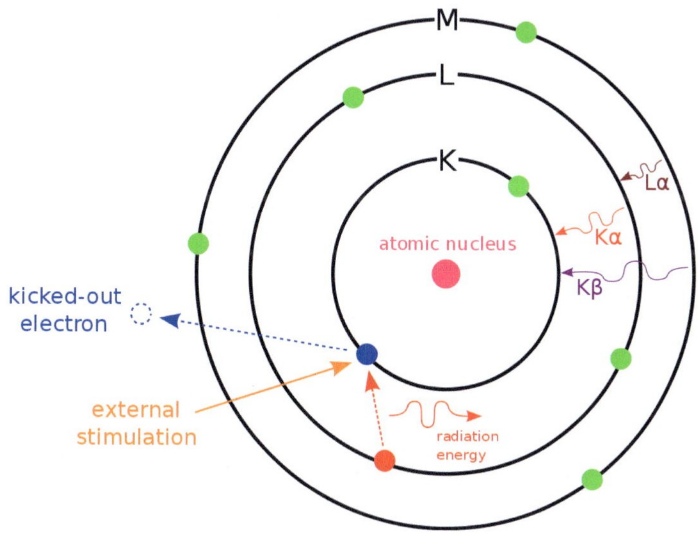

Figure 5.4. *Schematic of energy-dispersive X-ray spectroscopy (EDX).*

Source: Bobby Gaston, Creative Commons License.

The comprehensive coverage of evaluation of performance as well as quality control of synthetic components is provided. This article discusses several fundamental techniques for evaluating high-Tc superconductors' performance and quality. These techniques include electrical resistivity measurements, determination of superconducting transition temperature, critical current density measurements, and magnetic property characterization. The final section of the chapter pertains to the difficulties and prospective avenues in the domain of synthesis and fabrication methodologies. The statement underscores the importance of implementing methods that are both scalable and cost-effective, integrating diverse techniques, making progress in high-throughput synthesis, and exploring innovative approaches to augment the comprehension and utilization of high-temperature superconductors (Eckstein & Bozovic, 1995).

5.2. POWDER SYNTHESIS METHODS

5.2.1. Solid-State Reaction Method

The solid-state reaction technique is extensively employed for synthesizing high-temperature superconductors. The process entails the amalgamation and pulverization of the individual powders, succeeded by a heat treatment procedure conducted at elevated temperatures. The initial substances, commonly metal oxides or carbonates, are meticulously measured by the intended stoichiometric ratio of the ultimate superconductive compound.

The solid-state reaction commences by achieving homogeneity through the thorough mixing of powders. The desired outcome can be achieved through manual grinding utilizing an agate mortar or high-energy ball milling technique. The ball milling process facilitates effective mixing by applying kinetic energy to the powders, resulting in a decrease in particle size and an enhancement in homogeneity. The resultant powder amalgamation undergoes a thermal processing procedure following the mixing phase. The heat treatment process is commonly executed within a regulated environment, such as a continuous argon or oxygen flow, to avoid unwanted reactions or oxidation. The powder blend is subjected to a precise temperature and maintained for a specific period to facilitate the solid-state chemical transformations. This process enables the dispersion of atoms and the creation of the intended high Tc superconducting phase (Lange et al., 1995) (Figure 5.5).

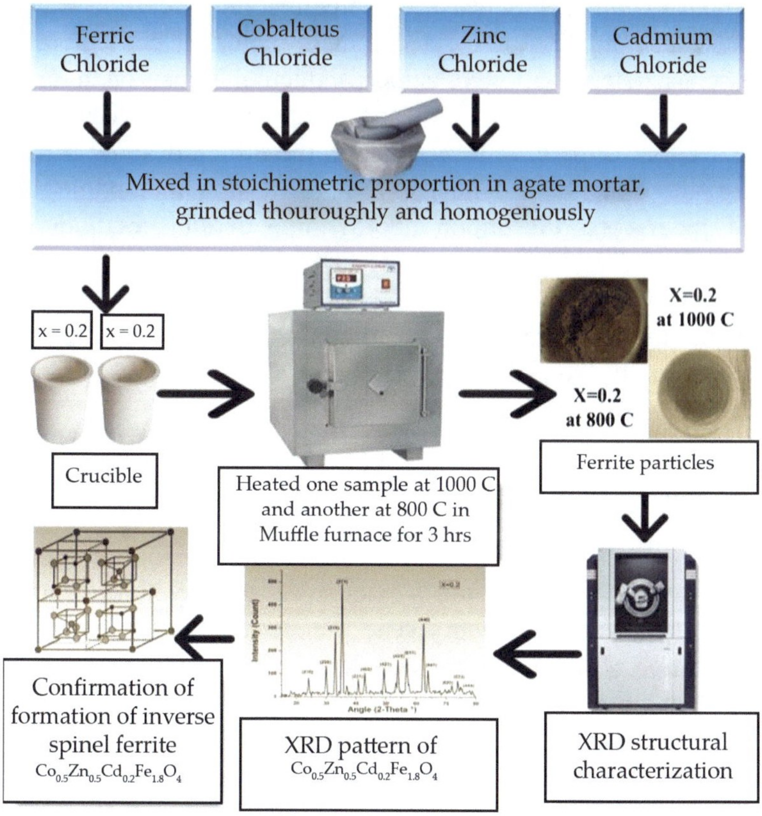

Figure 5.5. *Illustration of solid-state reaction method.*

Source: Shridhar Mathad, Creative Commons License.

The efficacy of the solid-state reaction technique is contingent upon meticulous management of the processing parameters, encompassing the heating rate, dwell time, and temperature profile. The above parameters significantly impact the ultimate material's phase purity, grain size, and superconducting characteristics. Additional procedures, such as grinding and annealing, can be implemented after the treatment to enhance the characteristics of the material further (Liew et al., 2000).

5.2.2. Ceramic Precursor Method

The ceramic precursor method, commonly called the PIT or "powder-in-tube" method, is frequently used in producing high-temperature superconductors. The process entails the initial production of a precursor powder, which is subsequently subjected to a regulated heat treatment to achieve the superconducting phase. The present methodology involves the dissolution of metal oxides or carbonates in a solvent appropriate to generate a precursor solution. The solvent selection is contingent upon various factors, including the solubility of the reactants and the intended reaction kinetics. Frequently employed solvents comprise aqueous solutions, alcohols, and solvents of organic origin (Marosi et al., 2002) (Figure 5.6).

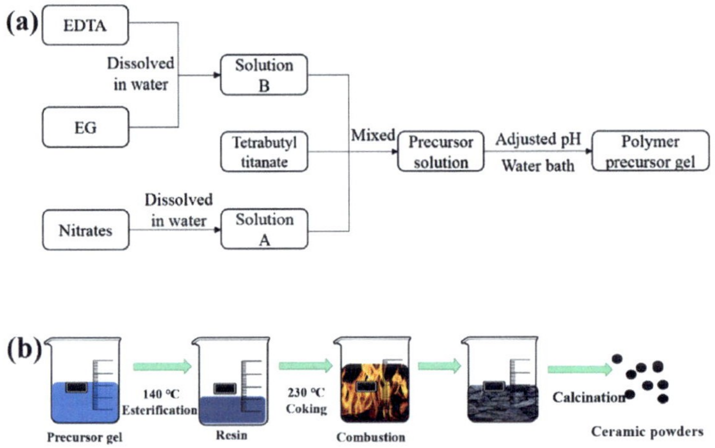

Figure 5.6. *Flow chart of preparation process of polymer precursor sol (a); and ceramic powders (b).*

Source: Yue Ming Li, Creative Commons License.

Following the preparation of the precursor solution, it is blended with a polymer binder, such as polyvinyl alcohol (PVA), to yield a thick slurry. The slurry undergoes additional processing through extrusion or doctor-blade casting to yield a pliable precursor tape or wire. The initial material in a tape or wire is commonly manipulated to achieve a specific shape, such as a planar or cylindrical wire. After its production, the precursor tape or wire undergoes a sequence of thermal processing stages to transform it into a substance that exhibits superconductivity. The first step involves subjecting the material to a drying process at low temperatures, which results in the solvent's removal and the binder's evaporation. This process ultimately yields a preform that is characterized by its porous nature. Subsequently, the preform undergoes a sequence of calcination and sintering procedures, wherein the precursor compounds undergo decomposition, reaction, and densification to yield the intended superconducting phase (Bhatia et al., 2002).

The process of calcination entails subjecting the preform to a moderate temperature to facilitate the elimination of any remaining organic compounds and encourage the development of intermediate phases. Subsequently, a sintering process is carried out, wherein the preform is subjected to elevated temperatures to promote grain coarsening and attain densification. The sintering procedure may be executed in diverse gaseous environments, including oxygen or argon, contingent upon the particular demands of the substance. The efficacy of the ceramic precursor technique is contingent upon the meticulous regulation of the thermal processing variables, encompassing the rate of temperature increase, duration of exposure, and thermal trajectory. The abovementioned parameters influence the ultimate material's phase purity, crystal structure, and superconducting characteristics. Using polymer binders facilitates the production of intricate geometries and the incorporation of superconducting substances into functional apparatus (Thakur & Singh, 2003).

5.2.3. Chemical Vapor Deposition (CVD)

Chemical Vapor Deposition (CVD) is a versatile technique extensively employed for synthesizing thin films and coatings, including high-temperature superconductors. The process entails the deposition of precursor gases onto a substrate, where they undergo chemical reactions to produce the intended superconducting material (Wang et al., 2013).

The chemical vapor deposition (CVD) technique commonly employs volatile metal-organic compounds or metal halides as precursors. The precursors are introduced into a reaction chamber and undergo thermal decomposition or reaction with other gases to produce the intended superconducting species. Subsequently, the resultant products of the reaction are deposited onto the substrate subjected to heat, forming a slender layer. The categorization of the CVD procedure is contingent upon the mode of precursor administration. In the context of conventional chemical vapor deposition (CVD), the precursor compounds are introduced into the reaction chamber in the presence of a carrier gas, such as hydrogen or nitrogen. The precursor vapors are conveyed to the heated substrate utilizing the carrier gas, where subsequent reactions occur (Hasenkox et al., 1998).

Metal-organic chemical vapor deposition (MOCVD) is a distinct subtype of cardiovascular disease (CVD) that employs volatile metal-organic precursors. The precursors mentioned above comprise metallic elements chemically bonded to organic ligands, including alkyl or alkoxides. The Metal-organic chemical vapor deposition (MOCVD) enables accurate manipulation of film composition through the modulation of precursor flow rates and reaction conditions. The Chemical Vapor Deposition (CVD) process necessitates meticulous regulation of multiple parameters, such as temperature, pressure, precursor concentration, and gas flow rates. These parameters influence the deposited film's growth rate, film morphology, crystallinity, and superconducting properties. The selection of the substrate material and its surface treatment are critical factors in attaining superior superconducting films (Carlsson & Martin, 2010).

Using chemical vapor deposition (CVD) presents numerous benefits in synthesizing high-temperature superconductors. The abovementioned technique facilitates meticulous regulation of the film's composition, thickness, and stoichiometry. The methodology mentioned above presents a remarkable degree of homogeneity across extensive surfaces. It is applicable for the deposition of thin films on a diverse range of substrates, such as single crystals, ceramics, and pliable substrates. Integrating superconducting materials into various device architectures facilitates the high-performance fabrication of electronic and energy devices (Nagel et al., 1982).

5.2.4. Sol-Gel Method

The sol-gel technique is a prevalent solution-based methodology utilized to produce high-temperature superconductors. The process entails creating a sol, a colloidal solution of nanoparticles, followed by gelation and subsequent heat treatment to achieve the intended superconducting phase (Dapkus, 1982).

The sol-gel methodology commences with the formulation of a precursor solution, accomplished by dissolving metal alkoxides or metal salts in an appropriate solvent. The solvent selection is contingent upon various factors, including the solubility of the reactants and the intended reaction kinetics. Frequently employed solvents encompass alcohols, namely ethanol or isopropanol, and aqueous solutions. After preparing, the precursor solution undergoes a regulated series of hydrolysis and condensation reactions, ultimately forming a sol. In hydrolysis, the metal alkoxides or metal salts undergo a reaction with water molecules, leading to the production of metal hydroxides or metal oxide clusters. Following this, condensation reactions take place, resulting in the creation of metal-oxygen-metal bonds and the enlargement of nanoparticles (Jasinski et al., 1987).

The colloidal solution comprises minute nanoparticles evenly distributed throughout the liquid medium. The manipulation of precursor composition, solvent properties, and reaction conditions can enable the regulation of nanoparticles' dimensions, morphology, and spatial arrangement. The sol is commonly subjected to a specific duration of aging to facilitate additional growth and stabilization of the nanoparticles. The introduction of a cross-linking agent or catalyst is necessary to induce the transformation of a sol into a solid gel. The introduction of a cross-linking agent facilitates the development of a tri-dimensional network architecture, converting a sol state into a gel-like substance. Subsequently, the gel undergoes a desiccation procedure to eliminate the solvent and achieve a porous solid. Subsequently, the desiccated gel is exposed to a heat treatment at elevated temperatures, commonly called calcination, to eliminate any residual organic constituents and facilitate the crystallization of the superconductive phase (Zhang et al., 2013).

The calcination process entails subjecting the gel to a predetermined temperature for a specific period. This stage facilitates the breakdown of precursor compounds, the elimination of organic residues, and the formation of the intended superconducting phase through crystallization. The control of calcination conditions, including heating rate, dwell time, and atmosphere, is crucial in ensuring the production of a superconducting material that is both pure and crystalline. The utilization of the sol-gel technique presents numerous benefits in the production of high-temperature superconductors. This technology facilitates the production of intricate geometries and thin coatings with meticulous manipulation of their composition and microstructural features. The procedure can be executed at comparatively lower temperatures, thereby mitigating the thermal strain on the materials. The process additionally permits the integration of dopants or modifiers to customize the characteristics of the superconductive substance, for instance, elevating the critical temperature or enhancing the flux-pinning capacity (Perednis & Gauckler, 2005).

5.2.5. Spray Pyrolysis

Aerosol deposition, spray coating, or spray pyrolysis make thin layers of high-temperature superconductors. Particles are broken into atoms, and precursor solutions are sprayed onto a heated substrate. The superconducting film forms during thermal breakdown and crystallization. A precursor solution of metal salts or metal-organic compounds is atomized into tiny droplets using a nebulizer or ultrasonic nozzle. The droplets then target a heated substrate at a preset temperature. The droplets evaporate quickly over the heated substrate, depositing a thin coating of the superconductive material (Mooney & Radding, 1982).

Controlling precursor content, droplet size, substrate temperature, and deposition rate optimizes film quality and crystallinity. The solvent and precursor composition affect film characteristics. Thermal treatment follows deposition. Thermal breakdown, crystallization, and phase transition are caused by heating the film for a certain time. Thermal treatment conditions are carefully chosen to ensure a clean, crystalline superconducting phase.

Spray pyrolysis produces high-temperature superconducting films with several advantages. This method deposits films on large-area surfaces and objects with complex shapes. The above approach achieves uniform coating thickness across large areas by manipulating deposition parameters. It also helps deposit films with numerous components and add dopants or modifiers to adjust superconductive properties (Patil, 1999).

5.3. SINGLE CRYSTAL GROWTH TECHNIQUES

5.3.1. Flux Method

The utilization of the flux method is a widely adopted approach for the cultivation of singular crystals of high-temperature superconductors. The present methodology involves the preparation of a compo1site comprising the intended superconducting material and flux material. The flux material functions as a solvent, thereby aiding the growth of singular crystals by creating a regulated milieu for crystallization (Messing et al., 1993).

The initial step involves subjecting the amalgamation of the superconductive substance and the flux substance to elevated temperatures surpassing the flux substance's melting point, resulting in a uniform liquefied state. Subsequently, the temperature is systematically reduced to facilitate the precipitation of the superconducting substance from the flux, thereby engendering the formation of solitary crystals. The deliberate deceleration of the cooling process is a pivotal factor in attaining single crystals of significant dimensions, exhibiting superior orientation and quality. The selection of the flux material is of utmost importance, as it must possess a melting point lower than the intended growth temperature and must not undergo any chemical reactions with the superconducting material. Bismuth oxide (Bi2O3), lead oxide (PbO), and thallium oxide (Tl2O3) are among the frequently utilized flux materials for high-temperature superconductors (Brown, 1988).

The flux method possesses a notable advantage in cultivating single crystals of substantial size. This characteristics holds significant relevance for various experimental investigations and pragmatic implementations. Including flux material mitigates impurities and flaws in the crystal lattice, enhancing superconducting characteristics. Prudent selection and optimizing the flux material are imperative to circumvent undesired chemical reactions or phase transitions (Koohpayeh, 2016).

5.3.2. Bridgman-Stockbarger Method

The vertical gradient freeze (VGF) method, commonly called the Bridgman-Stockbarger method, is a prevalent approach to growing single crystals encompassing high-temperature superconductors. The present methodology entails the utilization of a furnace that features a regulated temperature gradient running vertically (Figure 5.7).

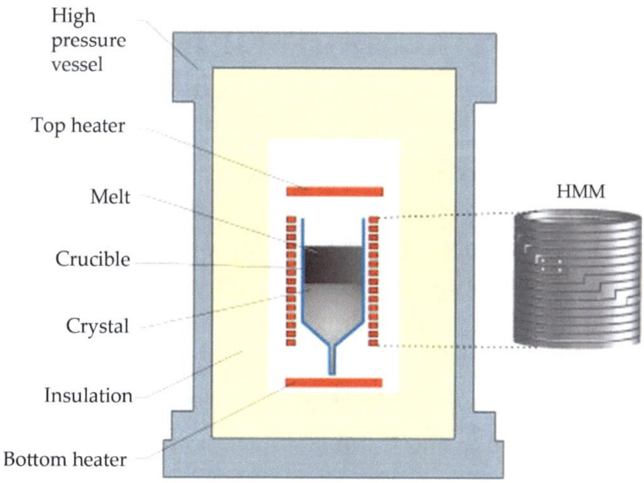

Figure 5.7. *Illustration of vertical gradient freeze (VGF) method.*

Source: Ch. Frank-Rotsch, Creative Commons License.

Remember

Post-processing and heat treatment steps, such as annealing and oxygenation processes, play a crucial role in optimizing the stoichiometry and doping levels of high-temperature superconductors.

The initial step involves placing the superconducting material, typically in the form of a polycrystalline ingot or powder, into a crucible. Subsequently, the crucible is subjected to a heating process to induce the melting of the material within the high-temperature region of the furnace.

As the crucible descends gradually through the temperature gradient, the liquefied substance undergoes solidification, forming a solitary crystal. The temperature gradient within the furnace significantly influences the crystal growth rate and quality determination. The solidification process must be meticulously regulated to facilitate a gradual and regulated solidification process, thereby enabling the crystal to grow in a regulated fashion with the intended crystallographic orientation (Kimura & Shindo, 1977).

The Bridgman-Stockbarger technique confers a notable benefit in its capacity to exert meticulous regulation over the growth parameters, including temperature gradient and pulling speed. Implementing this control leads to an enhancement in the quality of crystals and a decrease in defects. The technique necessitates meticulous optimization and regulation to circumvent complications such as thermal gradients, segregation, and impurity integration, which can harm the superconducting characteristics of the crystals (Abegunde et al., 2019).

5.3.3. Traveling Solvent Floating Zone Method

The floating zone method, also referred to as the traveling solvent floating zone method, is a versatile approach employed for the growth of single crystals of high-temperature superconductors. The technique entails utilizing a concentrated light source, such as a high-intensity halogen lamp or a laser, to selectively elevate the temperature of a limited area of the polycrystalline substance.

The application of concentrated thermal energy induces a localized phase transition, forming a liquified area. The manipulation of the feed and pulling rates of the crystal enables the molten zone to be displaced throughout the material with precision. The solidification of the molten zone leads to the formation of a solitary crystal during its traversal through the material. The crystal quality is refined, and the desired size is achieved through multiple process iterations (Iijima & Mastsumoto, 2000c).

The floating zone method with traveling solvent provides a high degree of control over the crystal growth process, enabling accurate manipulation of growth parameters. This process facilitates the generation of singular crystals that exhibit elevated levels of purity, minimized defects, and precisely defined crystallographic orientations. Utilizing a heat source confined to a specific region reduces thermal gradients and promotes uniformity in the crystal's composition. This approach is advantageous in cultivating singular crystals of intricate and demanding substances. The system can accommodate diverse compositions comprising multi-component systems and doped materials. The successful implementation of the traveling solvent floating zone technique necessitates meticulous optimization of multiple parameters such as temperature, pulling rate, and feed rate to guarantee the production of high-quality and defect-free crystals (Grosso et al., 2015).

5.3.4. Czochralski Method

The CZ method called the Czochralski method, is a commonly utilized approach for cultivating singular crystals of diverse materials encompassing high-temperature superconductors. The technique employed involves using a crucible capable of withstanding high temperatures, which contains the superconductive substance in either a polycrystalline rod or powder form (Figure 5.8).

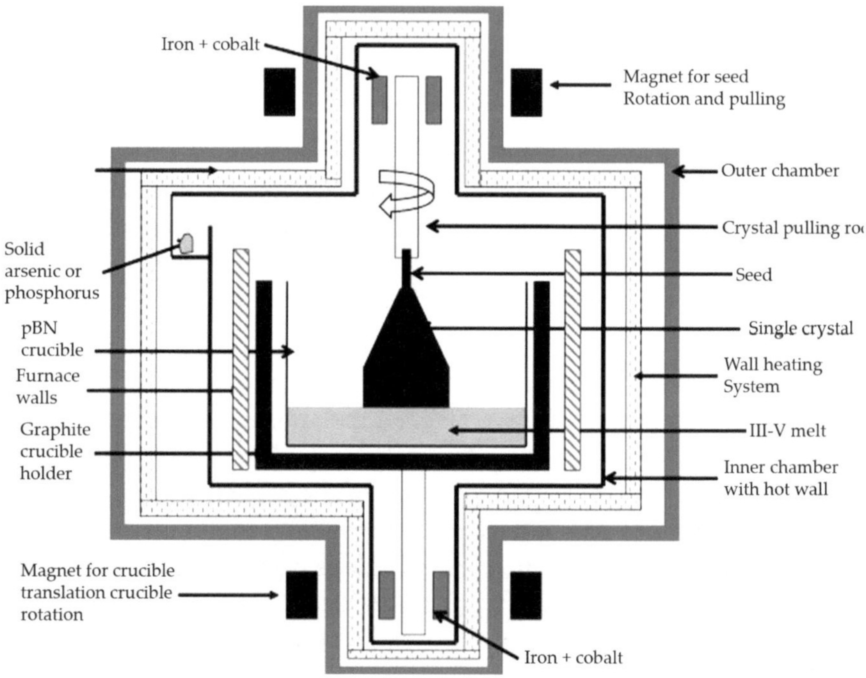

Figure 5.8. *Diagram of Czochralski method.*

Source: P. S. Dutta, Creative Commons License.

To commence the growth process, the crucible is subjected to heat to induce the melting of the material, resulting in the formation of a molten zone. Introducing a seed crystal commonly affixed to a rod into the molten substance and gradually withdrawing it while rotating is executed with great care. Upon the retraction of the seed crystal, the liquefied substance undergoes solidification in its vicinity, thereby facilitating the proliferation of a solitary crystal (Messier, 1988). Crystal rotation is an effective technique for preserving a consistent crystal structure and inhibiting the emergence of unfavorable defects. The Czochralski method presents various benefits in growing crystals with high-temperature superconducting properties. This technology facilitates the fabrication of single crystals of significant dimensions and superior quality while allowing for precise control over their crystallographic orientations. This technique facilitates the inclusion of dopants and accurately manipulates the crystal's composition. The Czochralski technique necessitates meticulous regulation of multiple variables, including temperature, pulling velocity, and rotation rate, to guarantee the production of flawless crystals and to reduce the inclusion of impurities. The procedure also necessitates meticulous oxygen level observation throughout the growth process to uphold the targeted oxygen stoichiometry, which is a critical factor in achieving high-temperature superconductivity (Hunt et al., 1993).

5.3.5. Molecular Beam Epitaxy (MBE)

Molecular beam epitaxy (MBE) is a specialized methodology to fabricate thin films and epitaxial layers of high-temperature superconductors. In contrast to the methods mentioned above for growing bulk crystals, molecular beam epitaxy (MBE) is a process that involves the precise deposition of atomic or molecular species onto a substrate within a vacuum setting. The process of Molecular Beam Epitaxy (MBE) involves the utilization of a high-vacuum chamber to facilitate the evaporation or sublimation of the source material. The resultant vapor then condenses onto a heated substrate, leading to the formation of a thin film. The deposition process takes place at the atomic or molecular level, thereby enabling meticulous regulation of the thickness and composition of the film. Effusion cells, also known as Knudsen cells, are utilized to evaporate or sublimate source materials (Martin, 1985).

The utilization of MBE presents various benefits in facilitating the expansion of elevated temperatures superconducting thin films. This technique offers superior regulation of the film's thickness, composition, and epitaxial characteristics. This methodology facilitates the expansion of intricate hetero-structures and meticulous regulation of doping levels. MBE enables the integration of diverse constituents and impurities with utmost accuracy at the atomic level (Oskam et al., 1998).

The successful implementation of MBE necessitates meticulous optimization of the deposition parameters, including but not limited to temperature, deposition rate, and flux ratio, to attain the intended film quality and superconducting characteristics. The requirement of an ultra-high vacuum environment in Molecular Beam Epitaxy (MBE) renders it a technically intricate and challenging method for growth (Tonelli et al., 2019).

5.4. THIN FILM DEPOSITION TECHNIQUES

5.4.1. Physical Vapor Deposition (PVD)

1. **Thermal Evaporation:** It is a common thin film PVD method. The source material or evaporation source is heated in a chamber without air molecules. The source material is usually placed in a crucible or refractory boat. Aluminum, gold, and silver are common supply materials.

   The source material vaporizes when heated, usually via resistive heating or electron beam bombardment. A substrate near the source material condenses the gaseous phase. The substrate is heated to improve film adherence and structure (Figure 5.9).

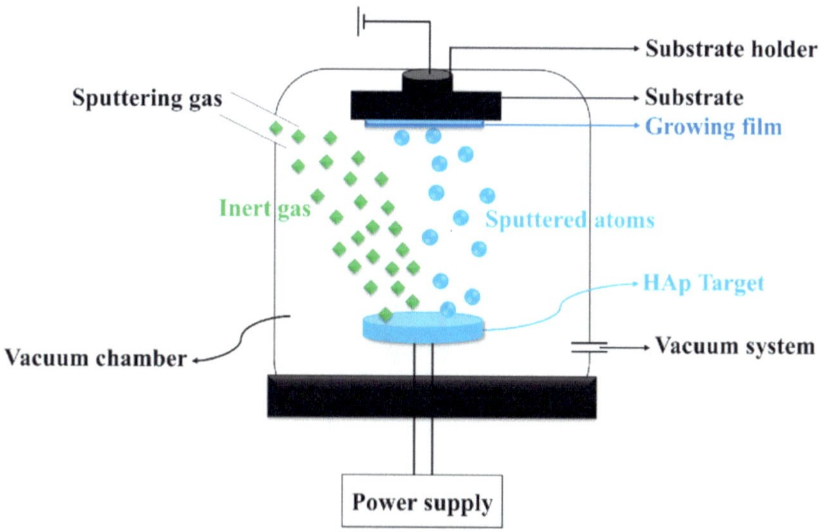

Figure 5.9. *Schematic of physical vapor deposition (PVD).*

Source: Mir Saman Savafi, Creative Commons License.

During condensation, evaporated material deposits in a linear pattern. Thermal evaporation's directionality makes film thickness and homogeneity controllable. Controlling heating temperature, deposition rate, and deposition duration yields the appropriate film properties.

Thermal evaporation works on metals, oxides, and organic molecules. Optoelectronics, photovoltaics, display technologies, and surface coatings use the material. However, materials with high melting points or low vapor pressures may need higher temperatures or other deposition methods (Shin et al., 2003).

2. **Sputtering:** It is a prevalent technique in physical vapor deposition utilized for the deposition of thin films. In contrast to thermal evaporation, sputtering is a process that entails using high-energy ions to bombard a solid target material, as opposed to relying on the heating of a source material. The ions are commonly produced via a plasma source inside a vacuum chamber.

   As a result of the collision between high-energy ions and the surface of the target, momentum is transferred to the atoms of the target material, leading to their displacement and subsequent ejection. The atoms that are expelled subsequently migrate toward a substrate and undergo condensation, ultimately resulting in the formation of a thin film. It is customary to place the substrate within the target's line of sight to ensure effective deposition.

   The sputtering technique presents various benefits, such as high-quality film purity, accurate regulation of film composition, and the capability to apply films on intricate three-dimensional substrates. This technique facilitates the deposition of materials with high melting points or low vapor pressures, which can prove arduous to accomplish using alternative deposition methodologies. Multiple sputtering techniques, such as direct current (DC) sputtering, radio frequency (RF) sputtering, magnetron sputtering, and reactive sputtering, exhibit unique benefits and utility (Golden et al., 1996).

3. **Molecular Beam Epitaxy (MBE):** It is a precise deposition technique utilized to cultivate thin films of high-quality crystals with accuracy at the atomic level. The process is conducted within an ultra-high vacuum environment to minimize the presence of contaminants and to guarantee accurate regulation of

the growth of the film. The utilization of MBE is prevalent in the manufacturing of semiconductor devices, optoelectronics and in the field of advanced materials research.

The process of molecular beam epitaxy (MBE) involves utilizing individual effusion cells or Knudsen cells to heat source materials, generating a molecular or atomic beam of the material. Subsequently, the beams comprising atoms or molecules are directed towards a substrate that has been heated, leading to their collision with the surface of the substrate. The collision-induced kinetic energy prompts the atoms or molecules to undergo deposition, forming a solitary crystal layer.

The controlled growth process in molecular beam epitaxy (MBE) involves precise manipulation of the substrate temperature and the evaporated material flux. The ability to exercise precise control facilitates the growth of films with well-defined characteristics such as thickness, composition, and crystalline structure. The MBE technique is renowned for its proficiency in producing exceptional-quality films characterized by atomically smooth surfaces and well-defined interfaces (Kumar et al., 2006).

The utilization of MBE presents various benefits, such as the capacity to cultivate heterostructures and superlattices, meticulous management over doping and alloy composition, and congruity with in-situ characterization methodologies. MBE is a multifaceted and costly methodology that necessitates considerable proficiency and advanced instrumentation for its execution.

Physical vapor deposition methods, including thermal evaporation, sputtering, and molecular beam epitaxy, offer flexible approaches to fabricating thin films with precise regulation of film thickness, composition, and structural characteristics. Diverse applications in fields such as electronics, optoelectronics, energy, and materials science necessitate the suitability of various techniques, each with its own set of advantages and limitations (Wu et al., 2018).

5.4.2. Chemical Vapor Deposition (CVD)

The process of Chemical Vapor Deposition (CVD) is a highly adaptable method for the deposition of thin films. It involves the deposition of films through the reaction of volatile precursor molecules on a substrate that has been heated. Chemical Vapor Deposition (CVD) operates on the fundamental principle of chemical reactions in the gaseous phase close to the surface of the substrate. The initial substances, commonly in a vaporous state, are inserted into a reactor vessel that is usually kept at a regulated temperature and pressure. To achieve a consistent distribution and regulate the composition of the film, it is common practice to dilute the precursor gases with a carrier gas.

Upon entering the reactor chamber, the precursor molecules undergo a series of chemical reactions. The reactions mentioned above may be instigated by thermal energy, plasma, or a synergistic interplay. The precursor compounds undergo decomposition, dissociation, or co-reactant-induced reactions, forming highly reactive species that can deposit onto the substrate's surface in the form of a solid film (Li et al., 2021).

The chemical vapor deposition (CVD) process comprises multiple stages for the deposition mechanism. Initially, the precursor molecules undergo adsorption onto the surface of the substrate that has been heated. The adsorption process is subject to various influences, including surface chemistry, temperature, and precursor concentration. Upon adsorption, the precursor molecules undergo surface reactions, wherein they undergo decomposition or react with other species to yield stable solid compounds. The chemical reactions in creating the desired film material may encompass bond-breaking, bond-forming, or atom rearrangement.

Chemical vapor deposition (CVD) presents various benefits compared to alternative deposition methods. A notable benefit of this technology is its exceptional ability to achieve conformal coating, enabling the uniform deposition of films on intricate three-dimensional configurations. The unique properties of CVD render it highly suitable for deployment in the realm of microelectronics, where there is a need for conformal coatings to be applied to intricate patterns and features (Liu et al., 2019).

In addition, Chemical Vapor Deposition (CVD) provides accurate manipulation of film thickness, consistency, and constitution. Customizing film properties to fulfill specific demands can be achieved by manipulating process parameters such as precursor flow rates, reactor temperature, pressure, and gas composition. Chemical vapor deposition (CVD) facilitates the deposition of various materials such as metals, metal oxides, nitrides, carbides, and diamond-like carbon.

Cardiovascular disease (CVD) has been utilized in diverse industrial and technological sectors. Chemical vapor deposition (CVD) is a prevalent technique in microelectronics for depositing dielectric materials, including silicon dioxide and silicon nitride, as well as conductive films, such as tungsten and aluminum. It also fabricates optical coatings, safeguarding coatings, and obstructive layers. Furthermore, Cardiovascular Disease (CVD) has acquired noteworthy importance in nascent domains, including the application of thin-film solar cells and the cultivation of graphene (Farges, 1982).

5.4.3. Atomic Layer Deposition (ALD)

Atomic Layer Deposition (ALD) deposits thin films with atomic-level control. The approach controls film thickness, homogeneity, and interface qualities, making it useful for microelectronics, energy-storage devices, and new technologies.

Sequential and self-limiting surface reactions drive ALD. Two or more precursor gases are introduced into a substrate-containing reaction chamber. Reactivity and thin film material production determine precursor gas choices. Precursor gases allow precise film composition and properties adjustment. ALD follows a pattern. The reaction chamber starts with the first precursor gas. The chemical adsorption of molecules forms a monolayer before the desired reaction on the substrate. The reaction self-limits to a single layer of atoms or molecules every cycle.

The chamber is cleaned after dosing the precursor. Purging the chamber with an inert gas like nitrogen or argon often eliminates it. Purging removes excess precursor and reaction by-products, preventing unwanted reactions in the following cycles. After purging, the precursor gas is injected, and the cycle is repeated. Each cycle deposits a regulated layer of material on the substrate. Cycles depend on film thickness, repeated until the appropriate thickness is reached (Shah et al., 2013).

ALD's features make it a tempting deposition approach. Its conformal coating capabilities are superb. Even on complex three-dimensional structures, the self-limiting reaction allows uniform film development. Atomic layer deposition (ALD) is ideal for coating trenches, vias, and nanoporous materials due to its conformality.

Atomic Layer Deposition (ALD) also controls film composition and interface properties. Precursor selection allows multi-component and multi-layered film deposition. To tailor film properties to device requirements, precise manipulation is needed. Microelectronics relies on Atomic Layer Deposition (ALD) for precise control and high-quality thin films. This method places gate dielectrics, diffusion barriers, and passivation layers in advanced semiconductor devices. ALD is used to make capacitors and batteries. To ensure performance and longevity, thin film coatings must have perfect thickness and homogeneity (Jin & Graebner, 1991). ALD is also important in nanoelectronics, photovoltaics, and flexible electronics. Atomically thin layers enable nanoscale research on novel materials and device architectures.

5.4.4. Electrochemical Deposition

Electrochemical deposition, also called electrodeposition or electroplating, is a multifaceted method employed to deposit thin films of metal or metal oxide onto a substrate. The process is based on the fundamental principles of electrochemistry, wherein an electrolyte solution containing metal ions is subjected to an electric current to facilitate the deposition process. In the coating process, the substrate functions as the cathode, whereas the anode comprises an electrode that shares the same metal as the ions in the electrolyte. The electrochemical deposition procedure commences with the formulation of an electrolytic solution. The solution commonly comprises metal salts or complexes that serve as a metal ion reservoir for deposition. The electrolyte selection is contingent upon the intended deposition of either a metal or metal oxide film. The deposition rate, film quality, and morphology can be affected by the composition of the electrolyte and its pH (Gambhir et al., 2015).

In the deposition process, the substrate is submerged in an electrolytic solution, and a direct or pulsed current is administered between the anode and cathode. During the flow of electric current, metal cations in the electrolyte migrate towards the cathode (also known as substrate), where reduction reactions occur. At the surface of the cathode, metallic ions undergo a reduction reaction by accepting electrons, forming a coherent metallic film. The efficacy of the reduction reaction is contingent upon the particular metal ions involved and the comprehensive electrochemical configuration of the cell. The electrochemical deposition process can be modulated by manipulating its parameters to regulate the properties of the resulting film. The applied voltage and current density influence the deposition rate and film thickness. The electrolyte composition and pH influence the composition and purity of the film. The duration of the deposition process is a critical factor in determining the thickness of the deposited film (Nayak et al., 2012).

The utilization of electrochemical deposition presents numerous benefits in producing thin films. The technique exhibits elevated deposition rates, rendering it appropriate for producing goods on a massive scale. The methodology mentioned above also facilitates exceptional conformal coverage, enabling coatings application onto intricate substrates with elevated aspect ratios. Furthermore, the procedure is comparatively uncomplicated and economical compared to alternative deposition methodologies. Electrochemical deposition finds its utility in diverse industrial sectors. Microelectronics employs conductive and barrier layers, such as copper interconnects, for deposition. Electroplating is a frequently utilized technique for applying ornamental coatings on metallic items and jewelry. The methodology mentioned above is also employed in the magnetic storage media manufacturing process, wherein slender layers of magnetic substances are coated onto substrates (Mishra et al., 2003).

Notably, various factors, including the bath temperature, agitation, and the existence of additives or surfactants in the electrolytic solution, can impact the electrochemical deposition procedure. Several factors have the potential to influence the quality, morphology, and adhesion of the film. Hence, meticulous regulation and enhancement of the deposition parameters are imperative to attain the intended film characteristics. To thoroughly comprehend electrochemical deposition, it is advisable to consult credible sources, scholarly articles, and specialized literature that delve into the particular considerations, applications, and advancements in this domain (Krebs, 2009).

5.5. *IN SITU* SYNTHESIS AND GROWTH TECHNIQUES

5.5.1. Pulsed Laser Deposition (PLD)

The Pulsed Laser Deposition (PLD) method is a versatile approach commonly employed for fabricating thin films exhibiting high-temperature superconductivity. In pulsed laser deposition (PLD), a laser pulse with high energy is focused onto a desired material, usually an oxide compound comprising the elements of interest. Upon application of the laser pulse, the target material undergoes vaporization, forming a plasma plume comprising atoms, ions, and clusters. Subsequently, the plasma plume undergoes expansion and condensation onto a substrate positioned close to the target. The process involves the condensation of atoms and ions present in the plasma plume, forming a thin film on the surface of the substrate (Veal et al., 1990).

The utilization of PLD presents various benefits in the production of high-temperature superconductors. The laser ablation technique accurately manipulates film composition, stoichiometry, and thickness. The elevated kinetic energy exhibited by the ablated species is conducive to favorable adhesion and crystallinity in the resultant film. The pulsed laser deposition (PLD) process can be conducted at comparatively low temperatures, which facilitates the formation of superconducting thin films on diverse substrates. Furthermore, the employment of PLD enables the real-time observation of the deposition process by utilizing methodologies such as reflection high-energy electron diffraction (RHEED) or spectroscopic ellipsometry (Mikelsen et al., 2005) (Figure 5.10).

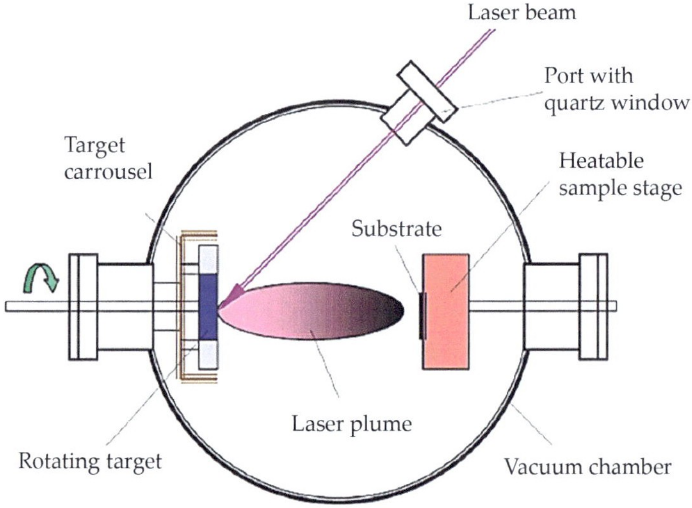

Figure 5.10. *Illustration of PLD device.*

Source: Vaccot, Creative Commons License.

5.5.2. Metal-Organic Chemical Vapor Deposition (MOCVD)

Metal-Organic Chemical Vapor Deposition (MOCVD) makes thin films with high-temperature superconductivity. A heated reactor chamber introduces volatile metal-organic precursors and carrier gas during MOCVD. Metal alkoxides and metal β-diketonates are used as precursors for element production. The reactor chamber thermally decomposes precursors, releasing metal atoms or complexes (Yao et al., 2012).

At high temperatures, metal atoms or complexes react with oxygen or water vapor to generate the desired superconducting material on a heated substrate. The substrate interface chemical reaction forms a thin layer. MOCVD accurately controls film chemical composition, thickness, and doping concentrations. Superconducting films may be made by optimizing precursor flow rates, temperature, and pressure (Feenstra et al., 1991).

5.5.3. Hybrid Physical-Chemical Vapor Deposition (HPCVD)

The HPCVD synthesis process combines PVD and CVD characteristics. HPCVD is ideal for growing thin films of high-temperature superconductors because it improves film characteristics and stoichiometry. HPCVD involves introducing a metal-organic precursor and a reactive gas like nitrogen or oxygen into a reactor chamber. The antecedent is usually a volatile metallic compound. The reactor chamber thermally decomposes the precursor, releasing metal atoms or complexes. After that, the metallic entities react with the high-reactivity gas to produce a thin layer on a heated substrate. The approach is called "hybrid" because it is affected by chemical and physical processes.

HPCVD produces high-quality, stoichiometric films and correctly regulates film properties. Precursor flow rates, temperature, and pressure can adjust film composition and crystallinity, enabling superconductivity. HPCVD allows physical and chemical systems to work together. This improves the film's durability and control over CVD and PVD (Kang et al., 2004).

5.5.4. Hybrid Molecular Beam Epitaxy (MBE)

Hybrid Molecular Beam Epitaxy (MBE) is a method that integrates aspects of molecular beam epitaxy and physical vapor deposition. The MBE technique is a widely recognized approach for fabricating thin films with high accuracy at the atomic level. Hybrid molecular beam epitaxy (MBE) is a potent technique for producing high-quality epitaxial films in high-temperature superconductors.

The hybrid molecular beam epitaxy (MBE) technique involves the introduction of a metal-organic precursor into an ultra-high vacuum (UHV) chamber in conjunction with additional molecular or atomic beams (Feng & Li, 2017).

In the ultra-high vacuum (UHV) environment, the precursor, commonly a volatile compound, experiences thermal decomposition upon heating, resulting in the liberation of metal atoms or complexes. The metallic species subsequently react with other beams or gases present in the chamber, such as oxygen or nitrogen, forming the intended superconducting thin film. Hybrid molecular beam epitaxy (MBE) precisely manipulates film composition, crystalline structure, and interface characteristics. The ultra-high vacuum (UHV) setting serves to mitigate the presence of contaminants and facilitates the cultivation of films possessing atomic-level smoothness. Regulating fluxes and temperatures with precision facilitates the development of intricate heterostructures and systems comprising multiple layers. The Hybrid Molecular Beam Epitaxy (MBE) technique has effectively produced high-temperature superconducting films, including YBCO and BSCCO, exhibiting remarkable crystalline characteristics and superconducting attributes (Rajamathi & Seshadri, 2002).

5.5.5. Electrochemical Deposition

Electrochemical deposition (electrodeposition) uses an electric field to deposit materials on a substrate. The approach synthesizes thin films and bulk-like high-temperature superconductors.

Electrochemical deposition dissolves superconducting material in an electrolyte solution, usually an acidic or alkaline bath with metal ions. The substrate is cathodic, and the conducting electrode is anodic. Metal cations in the electrolyte solution migrate toward the cathode (substrate) and are reduced at its surface when an electric current is applied. The reduction procedure creates a superconducting coating on the substrate. Electrochemical deposition helps high-Tc superconductor synthesis. Current density, bath composition, and deposition time may be adjusted to control film thickness, composition, and shape. Electrochemical deposition can mass-produce superconductive materials inexpensively. This approach is ideal for thick films and bulk-like materials. The electrochemical deposition has produced cuprate-based high-temperature superconductors like YBCO and BSCCO (Yu, 2001).

5.6. PROCESSING AND FABRICATION TECHNIQUES

1. **Mechanical Milling and Mixing:** Using mechanical milling and mixing methods is of paramount importance in synthesizing and fabricating high-temperature superconductors. The present procedure involves the mechanical milling and blending precursor powders to achieve uniformity and augment the materials' reactivity. The mechanical milling process entails utilizing high-energy ball mills or attritors, which subject the powders to rigorous mechanical forces. This leads to a decrease in particle size and the creation of composite powders that are finely dispersed. The

milling process facilitates the diffusion and intimate mixing of the constituents, thereby promoting the formation of a homogeneous precursor mixture. This methodology enables accurate regulation of composition and stoichiometry, resulting in enhanced superconducting characteristics. Furthermore, mechanical milling can initiate solid-state reactions and facilitate the creation of metastable phases, thereby broadening the scope of materials that can be produced (Li et al., 2016).

2. **Tape Casting and Thick Film Deposition:** High-temperature superconducting films and bulk components are made via tape casting and thick film deposition. Tape casting requires a steady slurry of superconductive particles, binding agents, solvents, and additives. Green tape is made by applying slurry to a flexible substrate or carrier film, drying, and sintering. Stacking, laminating, and cutting the green tape may create complicated geometries or multilayered constructions. Tape casting can produce large superconducting films with controllable thickness and uniformity in coated conductors and power devices.

   Screen printing and doctor blade coating produces thick superconducting films on many surfaces. Screen printing applies superconducting inks to a substrate using a stencil or screen, while doctor blade coating uses a blade to disperse the ink evenly. These approaches enable film thickness control and superconductive layers with high critical current densities. Superconducting wires, sensors, and microwave devices are made via thick film deposition (Allendorf et al., 2008).

3. **Laser Ablation and Patterning:** These techniques are imperative for the accurate and regulated production of extreme temperatures superconducting films and structures. The process of laser ablation entails the utilization of a laser beam with high energy to eliminate material from a specific target. This results in creating a plume consisting of vaporized material that subsequently deposits onto a substrate, forming a thin film. The methodology mentioned above facilitates the proliferation of superconductive epitaxial films that exhibit exceptional crystalline properties and interfaces of superior quality. Laser ablation presents a notable benefit of enabling selective area deposition, thereby facilitating the creation of intricate patterns and structures (Biz & Occelli, 1998).

Patterning methodologies, such as photolithography and ion beam etching, delineate the shape and size of superconducting films. The photolithography process entails utilizing light-sensitive photoresist materials that are selectively exposed to light through a patterned mask. Following exposure, either the exposed or unexposed regions are developed to generate a resist mask that can be employed for etching or deposition procedures. The process of ion beam etching involves the utilization of an ion beam to eliminate material from a superconducting film, adhering to a predetermined pattern established by a resist mask. Using patterning techniques facilitates the production of superconducting apparatus, such as Josephson junctions, resonators, and quantum circuits, with meticulous management of dimensions and alignment (Huang et al., 2011).

4. **Inkjet Printing and Additive Manufacturing:** The utilization of inkjet printing and additive manufacturing methods has garnered noteworthy interest in contemporary times for producing high-temperature superconductors. These techniques present benefits such as digital regulation, material effectiveness, and the capability to generate intricate shapes. The inkjet printing process entails the accurate placement of superconducting inks onto a substrate by utilizing printheads that operate via thermal, piezoelectric, or electrostatic mechanisms. The inks' composition comprises nanoparticles exhibiting superconductivity that is uniformly distributed in a fluid medium or a matrix that serves as a binding agent. Using inkjet printing technology enables the expeditious and precise application of superconducting trends or films while also allowing for scalability. The methodology mentioned above exhibits potential for utilization in various domains, including but not limited to flexible electronics, sensors, and fabrication as per requirement (Patra & Baek, 2015).

Additive manufacturing, also known as 3D printing, has surfaced as a propitious method for producing intricate three-dimensional configurations and operational apparatuses. Integrating high-temperature superconducting materials into printable pastes or filaments has facilitated the direct printing of superconducting components featuring complex geometries. The additive manufacturing process provides a range of design possibilities and enables the incorporation additional functional materials or components into the printed structures. Implementing this methodology can transform the manufacturing process of high-temperature superconducting apparatus, facilitating tailored configurations and expeditious prototyping (Singh, 2013).

5.7. POST-PROCESSING AND HEAT TREATMENT

5.7.1. Annealing and Oxygenation Processes

The annealing and oxygenation procedures significantly influence the post-processing and heat treatment of high-temperature superconductors (high-Tc). The annealing process entails the deliberate application of controlled heating and cooling cycles to the synthesized material, intending to improve its structural and superconducting characteristics. The oxygen annealing process is a prevalent technique utilized in materials science, whereby the specimen is subjected to a regulated oxygen environment at increased temperatures to achieve annealing. This procedure facilitates the optimization of the oxygen concentration, a critical factor in the superconducting properties of high-temperature superconducting materials (Blythe, 1984).

The annealing process entails structural modifications of the sample, which involve the creation of intended crystalline phases, removing imperfections, and enhancing grain interconnectivity. The regulated exposure to oxygen is a crucial factor in augmenting the oxygen content, facilitating the attainment of elevated critical temperatures, and amplifying the superconducting characteristics. Optimization of annealing conditions, including temperature, length of time, and oxygen partial pressure, is crucial for attaining the intended superconducting characteristics. The oxygenation processes primarily aim to incorporate oxygen into the lattice structure of high-Tc superconductors. Various techniques can be utilized to attain this objective, including oxygen diffusion from gaseous or solid origins, oxygen plasma processing, or oxygen annealing at elevated pressure. Accurate regulation of the oxygen concentration is of utmost importance due to its significant impact on both the superconducting characteristics and critical temperature. Through precise manipulation of the oxygen content via annealing and oxygenation techniques, scholars endeavor to optimize the superconducting transition temperature and augment the overall efficacy of high-Tc substances (McMillan, 1968).

5.7.2. Optimization of Stoichiometry and Doping Levels

Attaining the intended stoichiometry and doping levels is paramount in customizing the characteristics of elevated temperatures superconductors.

Stoichiometry pertains to the accurate atomic proportions of constituent elements within a given material. In contrast, doping entails deliberately incorporating contaminants or substituting atoms to alter the electronic and structural characteristics. The post-processing stage presents prospects for the optimization of stoichiometry and doping levels, which enables the precise adjustment of the material's superconducting properties.

Controlling stoichiometry pertains to the precise alignment of the atomic ratios of the high-Tc material with the desired composition. This objective can be attained through meticulous quantification of elemental ratios, blending elements, and utilizing compositional assessment methodologies such as X-ray fluorescence (XRF) or inductively coupled plasma (ICP) analysis. Modifying the stoichiometric ratios is a crucial step in mitigating the presence of surplus or inadequate constituents, which can significantly influence the crystal lattice, carrier density, and, ultimately, the superconducting characteristics. Doping is a deliberate process that entails introducing impurities or substituting atoms in high-Tc materials to modify their electronic and structural properties. The introduction of dopant atoms with varying valence states or electronic configurations can alter the concentration of charge carriers and impact the critical temperature. During the post-processing stage, there is a chance to meticulously regulate and enhance the doping levels by utilizing ion implantation, chemical vapor deposition, or solid-state reactions (Finkelshtein, 1996).

Through post-processing, researchers endeavor to optimize the superconducting characteristics, including but not limited to the critical temperature, critical current density, and vortex pinning, by adjusting the stoichiometry and doping levels. To carry out the optimization process effectively, a comprehensive comprehension of the phase diagram of the material, structure-property associations, and the impact of diverse dopants or contaminants on the electronic structure and superconducting behavior is necessary.

5.7.3. Quenching and Rapid Thermal Processing

Quenching and rapid thermal processing methods are implemented to prompt swift heating or cooling of high-temperature superconductors, consequently influencing their microstructural and phase stability attributes. The quenching process entails the expeditious reduction of temperature from elevated levels to ambient conditions, typically accomplished by submerging the specimen in a cooling agent, such as liquid nitrogen.

In contrast, rapid thermal processing refers to the expedited heating or reheating a given material to precise temperatures over brief intervals. Quenching is frequently employed to capture metastable phases or prevent undesired changes in phase that may arise during gradual cooling. As per the researchers ' findings, the preservation of a desired phase or metastable state through rapid quenching has resulted in distinct properties, such as heightened superconductivity. The quenching can impact the microstructure, grain size, and distribution of defects in a material, thereby influencing its electrical and magnetic properties (Werthamer, 1963).

Rapid thermal processing entails subjecting the sample to elevated temperatures for brief periods, usually from seconds to minutes. This methodology facilitates meticulous regulation of temperature profiles, thereby facilitating the examination of phase transformations, kinetics, and the impact of thermal treatment on the superconducting characteristics. Rapid heating or reheating can facilitate annealing, recrystallization, or grain growth, which in turn can result in increased interconnectivity between grains, decreased density of defects, and improved superconducting properties.

The attainment of the desired microstructural modifications in both quenching and rapid thermal processing techniques necessitates meticulous regulation of temperature, cooling rates, and duration. The techniques mentioned above present prospects for customizing the characteristics of the material through the manipulation of phase stability, defect structure, and grain boundaries, ultimately resulting in enhanced superconducting performance (Fleming et al., 1991).

5.7.4. Hydrothermal and Solvothermal Treatments

Hydrothermal and solvothermal treatments are post-processing techniques that specialize in subjecting superconductors with elevated temperatures to specific temperatures, pressure, and chemical environments through liquid solvents. The treatments mentioned above present distinctive benefits in regulating the substance's microstructure, crystal growth, and superconducting characteristics.

The hydrothermal treatment entails subjecting the specimen to high temperatures and pressures within a hermetically sealed vessel containing an aqueous solution or solvent. Under elevated pressure and temperature, the diffusion rates are increased, and chemical reactions are facilitated, thereby enabling the development of superior-quality single crystals or epitaxial thin films. The hydrothermal setting facilitates accurate regulation of crystal growth kinetics, morphology, and defect density, enhancing structural integrity and superconducting characteristics. The solvothermal treatment method is analogous to the technique; however, it utilizes organic solvents instead of aqueous solutions. Utilizing solvothermal conditions presents certain benefits, including an expanded selection of solvents, higher solubility of precursor materials, and the ability to regulate reaction kinetics. The implementation of solvothermal treatments has been observed to aid in producing intricate compositions, heterostructures, and nanoscale architectures. This presents a promising avenue for customizing superconducting properties at the nanoscale (Davar et al., 2009).

The attainment of desired crystalline structures, defect control, and superconducting properties necessitates meticulous optimization of temperature, pressure, solvent composition, and reaction time for hydrothermal and solvothermal treatments. Utilizing these methodologies presents distinctive opportunities for advancing superior monocrystals, slender layers, and nanoscale structures, thereby enabling scholars to investigate novel pathways for enhancing the efficacy of high-temperature superconductors.

5.8. CHARACTERIZATION OF SYNTHESIZED MATERIALS

5.8.1. X-Ray Diffraction (XRD)

X-ray diffraction (XRD) is a potent methodology employed in the structural analysis of synthesized materials within high-temperature superconductors. The material mentioned above is characterized by its structure of crystals, lattice parameters, and phase composition, which are all informative properties. The X-ray diffraction (XRD) technique operates based on the fundamental concept of constructive interference of X-rays scattered by the crystal lattice, leading to the emergence of distinct diffraction patterns characteristics of the sample. The crystallographic structure of a material can be determined through the analysis of its diffraction peaks. Furthermore, X-ray diffraction (XRD) can offer valuable information regarding the crystallinity level, contaminants' existence, and alterations in phases within the produced substance. This methodology is prevalent in characterizing both voluminous specimens and thin coatings of high-temperature superconductors, furnishing indispensable insights into the correlation between the material's structure and its superconductive characteristics (Li et al., 2010).

5.8.2. Scanning Electron Microscopy (SEM)

The utilization of Scanning Electron Microscopy (SEM) is a significant technique in the characterization of surface morphology and microstructure of synthetic substances exhibiting high-temperature superconductivity. The Scanning Electron Microscope (SEM) operates by utilizing a concentrated electron beam that traverses the surface of the specimen, thereby eliciting the emission of secondary electrons from the sample, which is then gathered. The images obtained from the process offer comprehensive insights into the topography, grain size, and sample distribution. Scanning electron microscopy (SEM) is a valuable tool for investigating high-temperature superconductors, as it allows for the identification of defects, grain boundaries, and interfaces. These features are known to have a significant impact on the electrical and magnetic properties of the material. Moreover, scanning electron microscopy (SEM), in conjunction with energy-dispersive X-ray spectroscopy (EDS), facilitates the examination of elemental properties, thereby facilitating the detection and cartography of chemical compositions present in the substance. The utilization of SEM is prevalent in optimizing synthesis parameters, evaluating sample quality, and comprehending the microstructure-property correlation in high-Tc superconductors (Ludwig et al., 2013).

5.8.3. Transmission Electron Microscopy (TEM)

The utilization of Transmission Electron Microscopy (TEM) is a prominent method for examining the atomic-level composition and imperfections in high-temperature superconducting materials that have been synthesized. This technique provides high-resolution imaging capabilities. Transmission electron microscopy (TEM) is a technique that employs a concentrated electron beam to traverse an extremely thin specimen, thereby enabling the observation of discrete molecules and crystal lattice configurations. The methodology mentioned above furnishes comprehensive insights into the material's crystal structure, encompassing the characteristics of grain boundaries, dislocations, and stacking faults. Moreover, Transmission Electron Microscopy (TEM) can detect the existence of nanoscale precipitates, phase separation, and imperfections that can substantially impact the superconducting characteristics. Using TEM's high-resolution imaging capabilities is crucial in comprehending the microstructure and defects present in high-Tc superconductors. This, in turn, facilitates the enhancement of synthesis techniques and optimization of superconductivity performance, rendering TEM an indispensable tool in this field (Petit et al., 1999).

5.8.4. Energy-Dispersive X-Ray Spectroscopy (EDX)

The utilization of Energy-dispersive X-ray spectroscopy (EDX) serves as a supplementary method to Scanning Electron Microscopy (SEM) and Transmission Electron Microscopy (TEM) for elemental analysis and mapping in the context of synthesized elevated temperatures superconducting materials. EDX utilizes the electron beam bombardment of material to quantify the emission of characteristics X-rays. Through the analysis of X-rays' energy and intensity, it is possible to ascertain the sample's elemental composition. The integration of Energy Dispersive X-ray Spectroscopy (EDX) with Scanning Electron Microscopy (SEM) or Transmission Electron Microscopy (TEM) is a common practice to acquire precise elemental data at specific locations within a sample. This technique facilitates the detection of contaminants, dopants, and disparities in chemical composition throughout the specimen. The utilization of this methodology is of great significance in ensuring quality control, exploring the phenomenon of elemental diffusion, and comprehending the impact of compositional fluctuations on the superconducting characteristics of high-Tc substances (Bianchi et al., 2001).

5.8.5. Raman Spectroscopy and Infrared Spectroscopy

Raman and infrared spectroscopy characterize produced high-temperature superconducting materials. The material's phonon modes and molecular vibrations provide vital information on its crystal symmetry, chemical bonding, and electronic characteristics. Raman spectroscopy measures the inelastic scattering of monochromatic light, while infrared spectroscopy measures sample absorption and reflection of infrared radiation. Both methods detect bond elongation, bending, and lattice oscillations. Raman and infrared spectroscopy can assess crystal quality, trace phase transitions, identify chemical species, and study the relationship between lattice vibrations and superconductivity in high-Tc materials. These non-invasive methods are ideal for in situ and operando superconducting equipment and thin coating research (Osuna et al., 1996).

Did You Know?

Raman spectroscopy and infrared spectroscopy are powerful characterization techniques used to analyze the vibrational and structural properties of synthesized high-temperature superconducting materials.

5.9. PERFORMANCE EVALUATION AND QUALITY CONTROL

5.9.1. Electrical Resistivity Measurements

Electrical resistivity measurements are essential for high-Tc superconductors' quality and performance evaluation. Electrical resistance is quantified by temperature. Resistivity data can reveal material conductivity and superconductivity initiation. This method lets researchers study the conductivity characteristics of high-temperature superconductors at different temperatures and find key factors.

Electrical resistivity tests require a thin film or bulk sample of the high-Tc superconductor. The specimen is then cooled using a cryostat or other means. After applying an electrical current, the four-point probe technique measures the voltage drop across the sample (Maranville et al., 2006).

Resistivity study yields important parameters. The resistivity transition width indicates superconductive material homogeneity and quality. A smaller transition width indicates a quick shift from normal conductivity to superconductivity, indicating high quality and purity. At the critical temperature (Tc), resistivity is zero. Superconductivity occurs at the critical temperature (Tc); hence its determination is crucial. The critical temperature (Tc) is essential for assessing high-temperature superconductors' suitability for various applications (Djerdj et al., 2007).

5.9.2. Superconducting Transition Temperature Determination

Assessing the superconducting transition temperature (Tc) is crucial in evaluating the efficacy and caliber of high-temperature superconductors. The critical temperature (Tc) denotes the temperature at which a material demonstrates a complete absence of electrical resistance and transitions into a superconducting state. Precise determination of Tc is of utmost importance in comprehending the characteristics of high-Tc substances and evaluating their appropriateness for pragmatic implementations (Huo & Wei,2009).

The four-point probe technique is a frequently employed approach for determining Tc. The proposed technique gradually cools the specimen while applying a low-level electric current. The voltage drop measurement across the sample is conducted by utilizing four electrodes strategically positioned at predetermined distances. The critical temperature can be identified by systematically manipulating the temperature and observing the point at which the resistance reaches zero. The methodology offers accurate and dependable evaluations of Tc and is extensively utilized in scientific investigation and quality assurance of high-Tc superconductors (Nelson et al., 1990).

Apart from the four-point probe method, various other techniques are employed to determine Tc precisely. The techniques encompassed in this category comprise AC susceptibility measurements and specific heat measurements. The AC susceptibility measurements process entails applying an alternating magnetic field to the specimen, followed by the measurement of the ensuing response. The abrupt alteration in magnetic susceptibility is utilized to detect the initiation of superconductivity. Measuring specific heat entails observing the material's heat capacity concerning temperature variation. The abrupt alteration in the specific heat observed at Tc denotes the shift from the conventional to the superconducting phase (Yang & Chen, 2004).

The precise determination of Tc enables researchers to evaluate the caliber and efficacy of superconducting materials with high Tc. This data is crucial in comprehending the conduct of the substances and enhancing their characteristics for particular uses.

5.9.3. Critical Current Density Measurements

Critical current density (Jc) is a key characteristics of elevated temperatures superconductors' performance and quality. "Jc" is the maximum current density a superconductor can handle without losing its superconductivity. Jc assessments provide insights into using high-Tc superconductors in power transmission, magnets, and fault current limiters.

J_c is measured using several experimental methods. Transport measurement involves applying an electrical current through the sample and measuring the voltage drop as a function of the magnetic field. Increasing the current and magnetic field strength, the voltage criteria, usually characterized by a threshold voltage, is reached, revealing the critical current density.

This method reveals the superconductor's magnetic field susceptibility and current limit (Plesch & Klimpel, 2002).

Magnetization measurements, which use a magnetic field to measure the specimen's magnetization, are also used to calculate Jc. Magnetization data may be used to calculate Hc and Jc. These data help explain how external magnetic fields affect high-Tc superconductors. They also provide crucial data for superconducting equipment and networks.

5.9.4. Magnetic Property Characterization

The assessment of magnetic properties plays a crucial role in the appraisal of performance and quality assurance of high-temperature superconductors. Comprehending the magnetic properties of said materials yields significant knowledge regarding their efficacy, essential characteristics, and appropriateness for diverse uses (Yang & Chen, 2004).

Magnetization measurements are a fundamental approach utilized in the characterization of magnetic properties. The methodology entails imposing an extrinsic magnetic field onto the superconductive specimen and quantifying the resultant magnetization. Crucial parameters can be ascertained by systematically manipulating temperature and magnetic field intensity. The previously mentioned parameters encompass the critical magnetic field (Hc), which denotes the uppermost magnetic field the superconductor can endure while retaining its superconducting state. The critical current density (Jc) is a crucial parameter that denotes the upper current limit that a superconductor can carry while retaining its superconducting properties (Plesch & Klimpel, 2002).

Magnetometry is a common technique employed to measure magnetization, often utilizing a magnetometer, such as a SQUID (Superconducting Quantum Interference Device) magnetometer. The specimen is exposed to diverse magnetic fields, and the resultant magnetization is quantified. The measurements yield data that can be utilized to ascertain crucial parameters, evaluate the superconductor's reaction to external magnetic fields, and appraise the material's quality (McCullough et al., 1976).

Apart from magnetization measurements, various other techniques are employed to characterize magnetic properties, including magnetic susceptibility measurements. A superconductor's magnetic susceptibility indicates its reaction to an externally applied magnetic field and its capacity to confine magnetic flux. The abovementioned measurements facilitate comprehension of the material's characteristics, its response to magnetic fields, and its possible utility in superconducting magnets and magnetic levitation systems (Iwata & Murai, 2006).

Activity 5.1

As a materials engineer, you are responsible for developing a new fabrication technique for high-temperature superconductors that improves their structural integrity. Create a step-by-step workshop where participants can learn and practice this innovative fabrication method.

5.10. SUMMARY

Chapter five provides an extensive overview of the synthesis and fabrication techniques employed in the production of high-temperature superconductors. It begins with an introduction to the chapter's topics and then explores various synthesis methods, including powder synthesis methods such as solid-state reaction, ceramic precursor, chemical vapor deposition (CVD), sol-gel, and spray pyrolysis. The chapter covers single crystal growth techniques, including the flux method, Bridgman-Stockbarger method, traveling solvent floating zone method, Czochralski method, and molecular beam epitaxy (MBE). It also delves into thin film deposition techniques, such as physical vapor deposition (PVD), chemical vapor deposition (CVD), atomic layer deposition (ALD), and electrochemical deposition. Furthermore, the chapter discusses in situ synthesis and growth techniques, processing and fabrication techniques, post-processing and heat treatment steps, and the characterization of synthesized materials using techniques such as X-ray diffraction (XRD), scanning electron microscopy (SEM), transmission electron microscopy (TEM), energy-dispersive X-ray spectroscopy (EDX), Raman spectroscopy, and infrared spectroscopy. Lastly, the chapter addresses the performance evaluation and quality control of high-temperature superconductors through electrical resistivity measurements, determining the superconducting transition temperature, critical current density measurements, and magnetic property characterization.

REVIEW QUESTIONS

1. Why is understanding synthesis and fabrication techniques crucial in the production of high-temperature superconductors?
2. Discuss the different powder synthesis methods used for high-temperature superconductors mentioned in this chapter.

3. What are the common single crystal growth techniques employed in the production of high-temperature superconductors?
4. Explain the advantages and applications of thin film deposition techniques in the field of high-temperature superconductors.
5. What are some in situ synthesis and growth techniques used for high-temperature superconductors?
6. Describe the processing and fabrication techniques involved in the production of high-temperature superconductors.
7. How do post-processing and heat treatment steps contribute to optimizing the properties of high-temperature superconductors?
8. Explain the importance of characterization techniques such as X-ray diffraction (XRD) and scanning electron microscopy (SEM) in analyzing synthesized materials.

REFERENCES

1. Abegunde, O. O., Akinlabi, E. T., Oladijo, O. P., Akinlabi, S., & Ude, A. U. (2019). Overview of thin film deposition techniques. *AIMS Materials Science*, *6*(2), 174–199.
2. Allendorf, M. D., Diver, R. B., Siegel, N. P., & Miller, J. E. (2008). Two-step water splitting using mixed-metal ferrites: thermodynamic analysis and characterization of synthesized materials. *Energy & Fuels*, *22*(6), 4115–4124.
3. Antonacopoulos, A., Bridson, D., Papadopoulos, C., & Pletschacher, S. (2009, July). A realistic dataset for performance evaluation of document layout analysis. In *2009 10th International Conference on Document Analysis and Recognition* (pp. 296–300). IEEE.
4. Baiman, S., & Demski, J. S. (1980). Economically optimal performance evaluation and control systems. *Journal of Accounting Research*, 184–220.
5. Behrman, E. C. (1987). Synthesis, characterization, and fabrication of high temperature superconducting oxides. *Advanced Ceramic Materials;(USA)*, *2*(CONF-8704372-).
6. Bhatia, T., Ozturk, A., Xie, L., Jordan, E. H., Cetegen, B. M., Gell, M., & Padture, N. P. (2002). Mechanisms of ceramic coating deposition in solution-precursor plasma spray. *Journal of Materials Research*, *17*(9), 2363–2372.
7. Bianchi, V., Caurant, D., Baffier, N., Belhomme, C., Chappel, E., Chouteau, G., & Wilmann, P. (2001). Synthesis, structural characterization and magnetic properties of quasistoichiometric LiNiO2. *Solid State Ionics*, *140*(1–2), 1–17.
8. Biz, S., & Occelli, M. L. (1998). Synthesis and characterization of mesostructured materials. *Catalysis Reviews*, *40*(3), 329–407.
9. Blythe, A. R. (1984). Electrical resistivity measurements of polymer materials. *Polymer Testing*, *4*(2–4), 195–209.
10. Brown, R. A. (1988). Theory of transport processes in single crystal growth from the melt. *AIChE Journal*, *34*(6), 881–911.
11. Carlsson, J. O., & Martin, P. M. (2010). Chemical vapor deposition. In *Handbook of Deposition Technologies for Films and Coatings,* William Andrew Publishing, *1*(1), (pp. 314–363).
12. Dapkus, P. D. (1982). Metalorganic chemical vapor deposition. *Annual Review of Materials Science*, *12*(1), 243–269.
13. Davar, F., Fereshteh, Z., & Salavati-Niasari, M. (2009). Nanoparticles Ni

and NiO: synthesis, characterization and magnetic properties. *Journal of Alloys and Compounds*, *476*(1–2), 797–801.

14. Djerdj, I., Arcon, D., Jagličić, Z., & Niederberger, M. (2007). Nonaqueous synthesis of manganese oxide nanoparticles, structural characterization, and magnetic properties. *The Journal of Physical Chemistry C, 111*(9), 3614–3623.
15. Eckstein, J. N., & Bozovic, I. (1995). High-temperature superconducting multilayers and heterostructures grown by atomic layer-by-layer molecular beam epitaxy. *Annual Review of Materials Science*, *25*(1), 679–709.
16. Edwards, S., Slattery, M. L., Mori, M., Berry, T. D., Caan, B. J., Palmer, P., & Potter, J. D. (1994). Objective system for interviewer performance evaluation for use in epidemiologic studies. *American Journal of Epidemiology*, *140*(11), 1020–1028.
17. Farges, J. P. (1982). A method for the "in-situ" synthesis and growth of indium phosphide in a Czochralski puller. *Journal of Crystal Growth*, *59*(3), 665–668.
18. Feenstra, R., Lindemer, T. B., Budai, J. D., & Galloway, M. D. (1991). Effect of oxygen pressure on the synthesis of $YBa2Cu3O_{7-x}$ thin films by post-deposition annealing. *Journal of Applied Physics*, *69*(9), 6569–6585.
19. Feng, S. H., & Li, G. H. (2017). Hydrothermal and solvothermal syntheses. In *Modern Inorganic Synthetic Chemistry* (pp. 73–104). Elsevier.
20. Finkel'shteĭn, A. M. (1996). Superconducting transition temperature in amorphous films. In *30 Years of the Landau Institute—Selected Papers* (pp. 288–291).
21. Fleming, R. M., Ramirez, A. P., Rosseinsky, M. J., Murphy, D. W., Haddon, R. C., Zahurak, S. M., & Makhija, A. V. (1991). Relation of structure and superconducting transition temperatures in A3C60. *Nature*, *352*(6338), 787–788.
22. Gambhir, S., Jalili, R., Officer, D. L., & Wallace, G. G. (2015). Chemically converted graphene: scalable chemistries to enable processing and fabrication. *NPG Asia Materials*, *7*(6), e186-e186.
23. Golden, T. D., Shumsky, M. G., Zhou, Y., VanderWerf, R. A., Van Leeuwen, R. A., & Switzer, J. A. (1996). Electrochemical deposition of copper (I) oxide films. *Chemistry of Materials*, *8*(10), 2499–2504.
24. Grosso, D., Boissière, C., & Faustini, M. (2015). Thin film deposition techniques. *The Sol-Gel Handbook*, 277–316.
25. Hasenkox, U., Hoffmann, S., & Waser, R. (1998). Influence of precursor chemistry on the formation of $MTiO_3$ (M= Ba, Sr) ceramic thin films.

Journal of Sol-Gel Science and Technology, *12*, 67–79.

26. Huang, X., Yin, Z., Wu, S., Qi, X., He, Q., Zhang, Q., & Zhang, H. (2011). Graphene-based materials: synthesis, characterization, properties, and applications. *Small*, *7*(14), 1876–1902.
27. Hunt, A. T., Carter, W. B., & Cochran Jr, J. K. (1993). Combustion chemical vapor deposition: A novel thin-film deposition technique. *Applied Physics Letters, 63*(2), 266–268.
28. Huo, J., & Wei, M. (2009). Characterization and magnetic properties of nanocrystalline nickel ferrite synthesized by hydrothermal method. *Materials Letters*, *63*(13–14), 1183–1184.
29. Iijima, Y., & Matsumoto, K. (2000). High-temperature-superconductor coated conductors: technical progress in Japan. *Superconductor Science and Technology*, *13*(1), 68.
30. Ikram, M., Raza, A., Altaf, S., Rafi, A. A., Naz, M., Ali, S., & Haider, J. (2021). High Temperature Superconductors. In *Transition Metal Compounds-Synthesis, Properties, and Application*, *1*(1), 1–20.
31. Ito, S., Hashizume, H., Yanagi, N., & Tamura, H. (2018). Advanced high-temperature superconducting magnet for fusion reactors: Segment fabrication and joint technique. *Fusion Engineering and Design*, *136*, 239–246.
32. Ito, S., Tamura, H., Yanagi, N., & Hashizume, H. (2021). Low-resistance joint development for segment-fabrication of high-temperature superconducting fusion magnets. *Nuclear Fusion*, *61*(11), 115002.
33. Iwata, M., & Murai, M. (2006). Buckling-restrained brace using steel mortar planks; performance evaluation as a hysteretic damper. *Earthquake Engineering & Structural Dynamics*, *35*(14), 1807–1826.
34. Jasinski, J. M., Meyerson, B. S., & Scott, B. A. (1987). Mechanistic studies of chemical vapor deposition. *Annual Review of Physical Chemistry*, *38*(1), 109–140.
35. Jin, S., & Graebner, J. E. (1991). Processing and fabrication techniques for bulk high-Tc superconductors: a critical review. *Materials Science and Engineering: B*, *7*(4), 243–260.
36. Kang, H. S., Kang, J. S., Kim, J. W., & Lee, S. Y. (2004). Annealing effect on the property of ultraviolet and green emissions of ZnO thin films. *Journal of Applied Physics*, *95*(3), 1246–1250.
37. Kimura, S., & Shindo, I. (1977). Single crystal growth of YIG by the floating zone method. *Journal of Crystal Growth*, *41*(2), 192–198.
38. Koohpayeh, S. M. (2016). Single crystal growth by the traveling solvent

technique: A review. *Progress in Crystal Growth and Characterization of Materials*, *62*(4), 22–34.

39. Krebs, F. C. (2009). Fabrication and processing of polymer solar cells: A review of printing and coating techniques. *Solar Energy Materials and solar cells*, *93*(4), 394–412.
40. Kumar, K., Ramamoorthy, K., Koinkar, P. M., Chandramohan, R., & Sankaranarayanan, K. (2006). A novel in situ synthesis and growth of ZnAl2O4 thin films. *Journal of Crystal Growth*, *289*(1), 405–407.
41. Lange, F. F., Tu, W. C., & Evans, A. G. (1995). Processing of damage-tolerant, oxidation-resistant ceramic matrix composites by a precursor infiltration and pyrolysis method. *Materials Science and Engineering: A*, *195*, 145–150.
42. Li, D. D., Tian, J. L., Gu, W., Liu, X., & Yan, S. P. (2010). A novel 1, 2, 4-triazole-based copper (II) complex: Synthesis, characterization, magnetic property and nuclease activity. *Journal of Inorganic Biochemistry*, *104*(2), 171–179.
43. Li, D., Adamo, C., Wang, B. Y., Yoon, H., Chen, Z., Hong, S. S., & Hwang, H. Y. (2021). Stabilization of $Sr_3Al_2O_6$ growth templates for ex situ synthesis of freestanding crystalline oxide membranes. *Nano Letters*, *21*(10), 4454–4460.
44. Li, J., Wu, Q., & Wu, J. (2016). Synthesis of nanoparticles via solvothermal and hydrothermal methods. In *Handbook of Nanoparticles* (pp. 295–328). Springer, Cham.
45. Liew, L., Zhang, W., An, L., Shah, S., Luo, R., Liu, Y., & Anseth, K. (2000). Ceramic MEMS. *American Ceramic Society Bulletin*, *80*(5), 25.
46. Liu, S. J., Han, S. D., Zhao, J. P., Xu, J., & Bu, X. H. (2019). In-situ synthesis of molecular magnetorefrigerant materials. *Coordination Chemistry Reviews*, *394*, 39–52.
47. Logvenov, G., Gozar, A., & Bozovic, I. (2009). High-temperature superconductivity in a single copper-oxygen plane. *Science*, *326*(5953), 699–702.
48. Ludwig, F., Eberbeck, D., Loewa, N., Steinhoff, U., Wawrzik, T., Schilling, M., & Trahms, L. (2013). Characterization of magnetic nanoparticle systems with respect to their magnetic particle imaging performance. *Biomedizinische Technik/Biomedical Engineering*, *58*(6), 535–545.
49. Maranville, B. B., McMichael, R. D., Kim, S. A., Johnson, W. L., Ross, C. A., & Cheng, J. Y. (2006). Characterization of magnetic properties at edges by edge-mode dynamics. *Journal of Applied Physics*, *99*(8), 08C703.

50. Marosi, G., Marton, A., Anna, P., Bertalan, G., Marosföi, B., & Szép, A. (2002). Ceramic precursor in flame retardant systems. *Polymer Degradation and Stability*, *77*(2), 259–265.

51. Martin, P. J. (1986). Ion-based methods for optical thin film deposition. *Journal of Materials Science*, *21*, 1–25.

52. McCullough, E. C., Payne, J. T., Baker Jr, H. L., Hattery, R. R., Sheedy, P. F., Stephens, D. H., & Gedgaudus, E. (1976). Performance evaluation and quality assurance of computed tomography scanners, with illustrations from the EMI, ACTA, and Delta scanners. *Radiology*, *120*(1), 173–188.

53. McMillan, W. L. (1968). Transition temperature of strong-coupled superconductors. *Physical Review*, *167*(2), 331.

54. Messier, R. (1988). Thin film deposition processes. *MRS Bulletin*, *13*(11), 18–21.

55. Messing, G. L., Zhang, S. C., & Jayanthi, G. V. (1993). Ceramic powder synthesis by spray pyrolysis. *Journal of the American Ceramic Society*, *76*(11), 2707–2726.

56. Mikelsen, M., Monakhov, E. V., Alfieri, G., Avset, B. S., & Svensson, B. G. (2005). Kinetics of divacancy annealing and divacancy-oxygen formation in oxygen-enriched high-purity silicon. *Physical Review B*, *72*(19), 195207.

57. Mishra, R. S., Ma, Z. Y., & Charit, I. (2003). Friction stir processing: a novel technique for fabrication of surface composite. *Materials Science and Engineering: A*, *341*(1–2), 307–310.

58. Mooney, J. B., & Radding, S. B. (1982). Spray pyrolysis processing. *Annual Review of Materials Science*, *12*(1), 81–101.

59. Nagel, S. R., MacChesney, J. B., & Walker, K. L. (1982). An overview of the modified chemical vapor deposition (MCVD) process and performance. *IEEE Transactions on Microwave Theory and Techniques*, *30*(4), 305–322.

60. Namburi, D. K., Shi, Y., & Cardwell, D. A. (2021). The processing and properties of bulk (RE) BCO high temperature superconductors: current status and future perspectives. *Superconductor Science and Technology*, *34*(5), 053002.

61. Nayak, R., Padhye, R., Kyratzis, I. L., Truong, Y. B., & Arnold, L. (2012). Recent advances in nanofiber fabrication techniques. *Textile Research Journal*, *82*(2), 129–147.

62. Nelson, S. B., Gardner, R. M., Crapo, R. O., & Jensen, R. L. (1990). Performance evaluation of contemporary spirometers. *Chest*, *97*(2), 288–297.

63. Oskam, G., Long, J. G., Natarajan, A., & Searson, P. C. (1998). Electrochemical deposition of metals onto silicon. *Journal of Physics D: Applied Physics*, *31*(16), 1927.
64. Osuna, J., de Caro, D., Amiens, C., Chaudret, B., Snoeck, E., Respaud, M., & Fert, A. (1996). Synthesis, characterization, and magnetic properties of cobalt nanoparticles from an organometallic precursor. *The Journal of Physical Chemistry*, *100*(35), 14571–14574.
65. Patil, P. S. (1999). Versatility of chemical spray pyrolysis technique. *Materials Chemistry and Physics*, *59*(3), 185–198.
66. Patra, J. K., & Baek, K. H. (2015). Green nanobiotechnology: factors affecting synthesis and characterization techniques. *Journal of Nanomaterials*, *2014*, 219–219.
67. Perednis, D., & Gauckler, L. J. (2005). Thin film deposition using spray pyrolysis. *Journal of Electroceramics*, *14*, 103–111.
68. Petit, C., Taleb, A., & Pileni, M. P. (1999). Cobalt nanosized particles organized in a 2D superlattice: synthesis, characterization, and magnetic properties. *The Journal of Physical Chemistry B*, *103*(11), 1805–1810.
69. Plesch, W., & Klimpel, P. (2002). Performance evaluation of the CoaguChek S system. *Haematologica*, *87*(5), 557–559.
70. Rajamathi, M., & Seshadri, R. (2002). Oxide and chalcogenide nanoparticles from hydrothermal/solvothermal reactions. *Current Opinion in Solid State and Materials Science*, *6*(4), 337–345.
71. Rehn, L. E. (1992). Ion beams in high-temperature superconductivity research. *Nuclear Instruments and Methods in Physics Research Section B: Beam Interactions with Materials and Atoms*, *64*(1–4), 161–168.
72. Rossnagel, S. M. (2003). Thin film deposition with physical vapor deposition and related technologies. *Journal of Vacuum Science & Technology A: Vacuum, Surfaces, and Films*, *21*(5), S74-S87.
73. Rupich, M. W., Li, X., Thieme, C., Sathyamurthy, S., Fleshler, S., Tucker, D., & Slack, J. (2009). Advances in second generation high temperature superconducting wire manufacturing and R&D at American Superconductor Corporation. *Superconductor Science and Technology*, *23*(1), 014015.
74. Rupich, M. W., Verebelyi, D. T., Zhang, W., Kodenkandath, T., & Li, X. (2004). Metalorganic deposition of YBCO films for second-generation high-temperature superconductor wires. *MRS bulletin*, *29*(8), 572–578.
75. Sengupta, S. (1998). High-temperature superconductors: Synthesis techniques and application requirements. *Journal of Minerals, Metals and*

Materials Society, *50*(10), 19–26.

76. Shah, M., Kwon, H. T., Tran, V., Sachdeva, S., & Jeong, H. K. (2013). One step in situ synthesis of supported zeolitic imidazolate framework ZIF-8 membranes: Role of sodium formate. *Microporous and Mesoporous Materials*, *165*, 63–69.
77. Shin, H. C., Dong, J., & Liu, M. (2003). Nanoporous structures prepared by an electrochemical deposition process. *Advanced Materials*, *15*(19), 1610–1614.
78. Singh, Y. (2013). Electrical resistivity measurements: a review. In *International Journal of Modern Physics: Conference Series* (Vol. 22, pp. 745–756). World Scientific Publishing Company.
79. Strauss, K., Hannet, I., Engels, S., Shiba, A., Ward, D. M., Ullery, S., & Kestens, L. (1996). Performance evaluation of the FACS Count system: a dedicated system for clinical cellular analysis. *Cytometry: The Journal of the International Society for Analytical Cytology*, *26*(1), 52–59.
80. Thakur, A., & Singh, M. (2003). Preparation and characterization of nanosize Mn0.4Zn0.6 Fe_2O_4 ferrite by citrate precursor method. *Ceramics International*, *29*(5), 505–511.
81. Tonelli, D., Scavetta, E., & Gualandi, I. (2019). Electrochemical deposition of nanomaterials for electrochemical sensing. *Sensors*, *19*(5), 1186.
82. Veal, B. W., You, H., Paulikas, A. P., Shi, H., Fang, Y., & Downey, J. W. (1990). Time-dependent superconducting behavior of oxygen-deficient YBa_2Cu_3Ox: Possible annealing of oxygen vacancies at 300 K. *Physical Review B*, *42*(7), 4770.
83. Wang, S. F., Zhang, J., Luo, D. W., Gu, F., Tang, D. Y., Dong, Z. L., & Kong, L. B. (2013). Transparent ceramics: Processing, materials and applications. *Progress in Solid State Chemistry*, *41*(1–2), 20–54.
84. Weinstein, R., Chen, I. G., Liu, J., & Lau, K. (1991). Permanent magnets composed of high temperature superconductors. *Journal of Applied Physics*, *70*(10), 6501–6503.
85. Werthamer, N. R. (1963). Theory of the superconducting transition temperature and energy gap function of superposed metal films. *Physical Review*, *132*(6), 2440.
86. Wu, Y., Li, D., Zhou, F., Liang, H., Liu, Y., Hou, W., & Tan, W. (2018). Versatile in situ synthesis of MnO_2 nanolayers on upconversion nanoparticles and their application in activatable fluorescence and MRI imaging. *Chemical Science*, *9*(24), 5427–5434.
87. Yang, C. C., & Chen, B. S. (2004). Key quality performance evaluation

using fuzzy AHP. *Journal of the Chinese Institute of Industrial Engineers*, *21*(6), 543–550.

88. Yao, Z., Nie, H., Yang, Z., Zhou, X., Liu, Z., & Huang, S. (2012). Catalyst-free synthesis of iodine-doped graphene via a facile thermal annealing process and its use for electrocatalytic oxygen reduction in an alkaline medium. *Chemical Communications*, *48*(7), 1027–1029.
89. Yu, S. H. (2001). Hydrothermal/solvothermal processing of advanced ceramic materials. *Journal of the Ceramic Society of Japan*, *109*(1269), S65-S75.
90. Zahi, S. (2010). Nickel–zinc ferrite fabricated by sol–gel route and application in high-temperature superconducting magnetic energy storage for voltage sag solving. *Materials & Design*, *31*(4), 1848–1853.
91. Zhang, Y. I., Zhang, L., & Zhou, C. (2013). Review of chemical vapor deposition of graphene and related applications. *Accounts of Chemical Research*, *46*(10), 2329–2339.

CHAPTER

6

Crystal Structures of Cuprate Superconductors

Contents

6.1. INTRODUCTION

We have already noticed that layered crystal structures are a characteristics property of all cuprate superconductors. Electrically conducting copper oxide layers are sandwiched between insulating layers, which act as charge carrier reservoirs. In this chapter, we will consider the crystal structures of the cuprate superconductors in detail (Park & Snyder, 1995).

The crystal structure of solids can be studied by different diffraction methods. In Section 6.2, X-ray and neutron diffraction will be briefly described. The wavelength of the X-rays or the particle waves must be comparable to or smaller than the lattice parameters of the investigated solids. Deceleration of 40kV electrons by a copper target generates bremsstrahlung with wavelengths in the region of interest. The wavelength λ of neutrons is given by the de Broglie relation $\lambda = h/p$, where h is Planck's constant and p the momentum of the particle. X-rays, mainly interacting with the electrons of atoms, are only weakly scattered by light elements. It is therefore difficult to study the occupation of lattice sites by oxygen in the cuprate superconductors by means of X-ray diffraction. More information on the positions of oxygen atoms in high-temperature superconductors can be obtained by neutron diffraction, which is more sensitive to the detection of light atoms. Due to their magnetic moment, neutrons can be used to study the magnetic structure of a solid. Neutron scattering performed below the Néel temperature (T_N) responds to the magnetic unit cell of the antiferromagnetic state, whereas diffraction patterns taken above T_N provide information on the atomic positions, which can be compared with that derived from X-ray diffraction (Tsukada et al., 2005).

In addition, X-ray powder diffraction can be used to identify the different solid phases present in a sample. The X-ray powder diffraction pattern of a single-phase polycrystalline sample is like a fingerprint characteristics of the solid phase in question. In multiphase samples, the ratios of the intensities of the strongest reflections of the different phases present can be used to estimate the fractions of these phases present in the sample. X-ray powder diffraction is therefore a powerful tool to find the optimum processing conditions to synthesize single-phase cuprate superconductors. In Section 6.3, the crystal structures of various cuprate superconductors are described. In addition, typical X-ray powder diffraction patterns of various cuprate superconductors will be presented (Axe & Crawford, 1994).

6.2. DIFFRACTION METHODS

6.2.1. Bragg Condition

A small fraction of the incident beam is reflected by each of the parallel lattice planes. Similar to the reflection of light at a mirror, the angles of incidence and reflection at each of the lattice planes are equal. Constructive interference of the beams, reflected from adjacent lattice planes, occurs for path differences of $n \cdot \lambda$, where n is an integer number. Thus, the propagation directions of scattered waves satisfy the Bragg condition

$$2d \sin\theta = n\lambda \qquad (1)$$

In this elastic scattering process, the energy of the X-ray photons is unchanged (Birkl et al., 1995).

6.2.2. Miller Indices

The different planes in the crystal lattice are described by the Miller indices. They can be found by the following procedure. First, the intercepts of the lattice plane with the crystallographic axes are determined. The Miller indices, typically called (*hkl*), are the smallest possible integer number ratios of the inverse values of these intercepts.

Tip

Understanding diffraction methods, such as X-ray diffraction and neutron diffraction, is essential for analyzing the crystal structures of cuprate superconductors.

In the case of the trigonal and hexagonal crystal systems, the Miller indices are frequently replaced by the Miller–Bravais indices (*hkil*). The values of h, k, and l are identical to those of the corresponding Miller indices. The extra index (i) is defined for convenience to be $i = -(h + k)$ (Sun et al., 2020) (Figures 6.1 and 6.2).

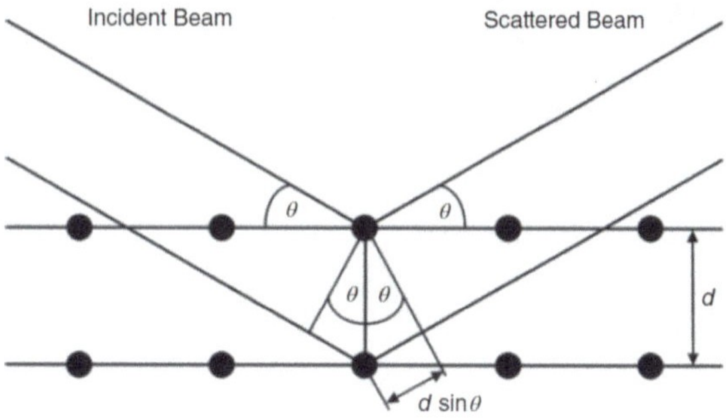

Figure 6.1. *Sketch of Bragg scattering. Path differences equal to an integer number of the wavelength of the X-rays lead to constructive interference.*

Source: Hydrargyrum, Creative Commons License.

Remember

Exploring the crystal structures of cuprate high-temperature superconductors, including La2CuO4, Bi-22(n–1)n compounds, Tl-based superconductors, and Hg-based superconductors, provides insights into their unique structural properties and superconducting behavior.

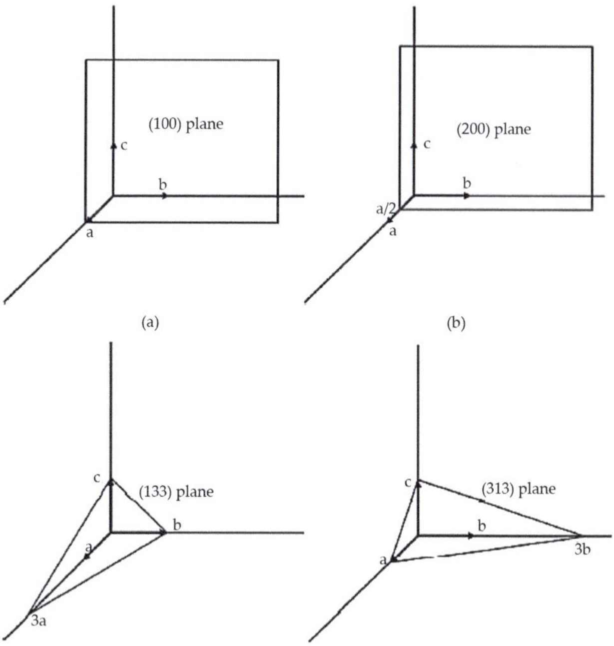

Figure 6.2. *Selected lattice planes and corresponding Miller indices (hkl).*

Source: Roger Nix, Creative Commons License.

6.2.3. Classification of Crystal Structures

Crystal structures are classified according to their translational symmetry. Fourteen Bravais lattices exist in 3D space, and only five on a 2D plane. The Bravais lattice symbols indicate both the crystal system and the centering type. The centering types in three dimensions and the corresponding symbols are primitive (*P*), face-centered (*C* (*A*, *B*)), body-centered (*I*), and all face-centered (*F*). In two dimensions, there exist only two centering types, namely primitive (*p*) and centered (*c*). For example, the lattice symbol for a face-centered cubic crystal structure is *cF* (Ziletti et al., 2018).

Based on their symmetry properties, all crystal structures can be classified into 230 space groups described by the Mauguin–Hermann symbols. For example, in the space group symbol *I4/mmm*, the letter *I* stands for body centering, the number *4* indicates a fourfold rotational symmetry, and the three letters *m* represent three mirror planes. We shall see later that many cuprate superconductors crystallize in the symmetry of the space group *I4/mmm* (Park et al., 2017).

6.2.4. X-Ray Diffraction

For the determination of the crystal structure of a solid by means of X-ray diffraction, single crystals as well as powder samples can be used. The Laue method, uses a single crystal and needs a continuous X-ray spectrum. The incident beam is parallel to one of the principal crystal directions. The positions of the diffraction spots obey the Bragg condition.

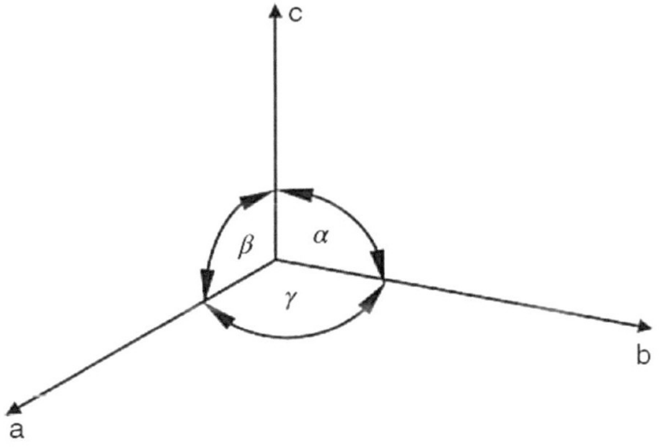

Figure 6.3. *Definition of the angles between the crystallographic axes.*

Source: Zbigniew Tylczynski, Creative Commons License.

For each crystallographic plane, the wavelength appropriate to fulfill the Bragg condition is selected out of the continuous X-ray spectrum. The pattern of the diffraction spots reveals the symmetry of the crystal lattice in question. The main disadvantage of the Laue method is the fact that single crystals of sufficient size and quality are required for the determination of the crystal structure. Frequently, suitable crystals are not available immediately after the discovery of a new superconductor (Patterson, 1944) (Figures 6.3 and 6.4; Tables 6.1 and 6.2).

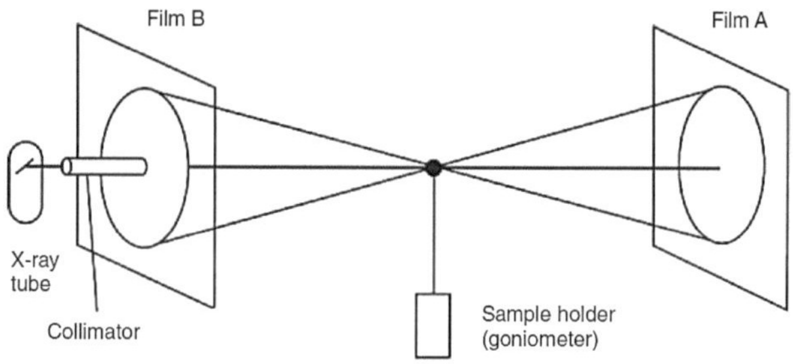

Figure 6.4. *Sketch of the Laue method.*

Source: Ruma Arora, Creative Commons License.

As an alternative to the Laue method, X-ray powder diffraction can be used to determine the crystal structure. The experimental setup of X-ray powder diffraction is sketched. A monochromatic X-ray beam is required for powder diffraction experiments. The characteristics X-rays produced by the excitation of inner electrons may be used as a source of monochromatic X-rays. Cu Kαradiation with a wavelength of 0.15418nm is frequently used for X-ray powder diffraction; nickel filters are then used to absorb most of the Kβline and the undesirable bremsstrahlung. Alternatively, a mono chromator crystal may be used, which reflects only the Cu Kαradiation. The position of the X-ray tube is fixed. The width of the incident X-ray beam is limited by the collimators C1 and C2. The sample holder is rotated around its axis, starting from the angle $\theta i$with a step width $\Delta\theta$, up to the final Bragg angle θ_f. Simultaneously, the angle between the detector and the incident beam is varied, starting from $2\theta_i$with a step width of $2\Delta\theta$, up to the final scattering angle of $2\theta_f$. The width of the reflected beam is limited by the collimators C3 and C4. The nickel filter, used to absorb most of the Kβline and the undesired bremsstrahlung background, is positioned between the collimators C3 and C4. The positions of the diffracted X-ray beams satisfy the Bragg condition. The intensities of the scattered X-rays are plotted as a function of the scattering angle 2θ. In addition to the determination of the crystal structure of a newly discovered superconductor, impurity phases present can be detected by X-ray powder diffraction (Takita et al., 1988) (Figure 6.5).

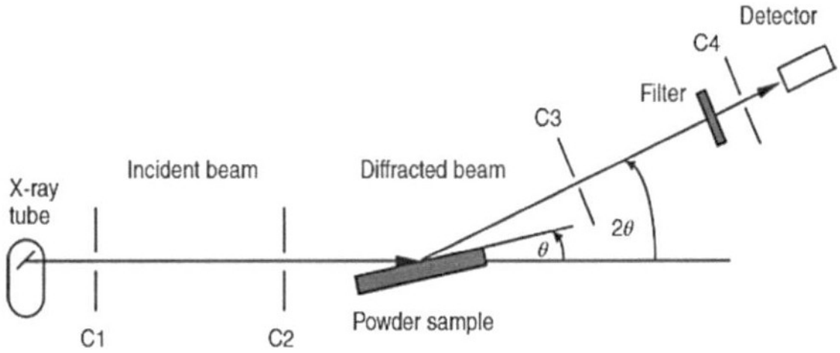

Figure 6.5. *Sketch of X-ray powder diffraction.*

Source: Nice CXone Expert, Creative Commons License.

The geometrical theory of diffraction described so far is sufficient to determine the lattice constants of the sample. The observed intensities of the scattered beams provide additional information about the occupation of the lattice sites. This information can be extracted by a comparison of calculated and measured intensities. The intensity of the diffracted X-rays is proportional to the structure factor $F(hkl)$ (Kudoh et al., 1985).

$$F(hkl) = \sum_j f_j \exp(-i2\pi(x_j h + y_j k + z_j l)) \quad (2)$$

where; the summation is over all atoms in the unit cell, f_j is the atomic form factor, $i^2 = -1$, h, k, l are the Miller indices, and the coordinates x_j, y_j, and z_j give the position of the jth atom in the unit cell. The atomic form factor is an atomic property mainly determined by the local electron concentration (Cao & Tan 2005).

6.2.5. Neutron Diffraction

The neutrons and the atomic nuclei interact via the strong interaction. The scattering of neutrons at the atomic nuclei in a solid provides in principle the same information as X-ray diffraction. The X-rays interact with the electrons of the atoms, and hence their scattering depends strongly on the atomic number (proton number). X-rays are only weakly scattered by light atoms, whereas they are strongly scattered by heavy atoms with a large atomic number. In contrast, neutron scattering depends only weakly on the atomic mass, and hence can provide information on the positions of light atoms in the lattice. Furthermore, the interaction of the magnetic moment of the neutron with the ordered magnetic moments, for example in an antiferromagnetic solid, provides information on the magnetic unit cell of the solid. In antiferromagnetic solids, the magnetic structure is different from the crystal structure. Frequently, the magnetic moment of adjacent atoms changes its sign, and hence the distance of magnetically identical atoms is larger than that of crystallographic identical atoms (O'connor & Valentine, 1969; Dunlop, 1973; Vaz, 2014).

The wavelength of neutrons is given by the formula $\lambda = 0.0286/E^{0.5}$, where λ is the wavelength of the neutrons in nanometers and E their kinetic energy in electron-volt (eV). An energy of 80meV leads to a wavelength of 0.1nm, which is comparable to the interatomic distances in a solid. Because of the low energy of the neutrons, inelastic scattering can be used to study excitations (e.g., phonons, magnons) in a solid. As an example, let us consider the generation or destruction of a phonon in an inelastic scattering process. For inelastic scattering, the Bragg condition can be written as (Salmon & Lond, 1992; Dong, 2018).

$$\mathbf{k} = \mathbf{k}' + \mathbf{G}_{\mathbf{hkl}} \pm \mathbf{q} \quad (3)$$

where; $\mathbf{k}$ is the wave vector of the incident neutron beam; $\mathbf{k}'$ the wave vector of the scattered beam; $\mathbf{G}_{\mathbf{hkl}}$ a reciprocal lattice vector; and $\mathbf{q}$ the wave vector of the phonon. The plus sign describes the generation of a phonon, while the minus sign corresponds to the destruction of a phonon. In addition, energy must be conserved in the scattering process (Nandi et al., 2016; Dyson, & McIndoe, 2000).

$$\frac{\hbar^2 k^2}{2m_n} = \frac{\hbar^2 k'^2}{2m_n} \pm \hbar\omega(q) \quad (4)$$

here; ω is the angular frequency of the phonon (Staron, 2017; Betzel et al., 1984).

6.3. CRYSTAL STRUCTURES OF THE CUPRATE HIGH-TEMPERATURE SUPERCONDUCTORS

6.3.1. The Crystal Structure of La_2CuO_4

The crystal structures of many cuprate high-temperature superconductors have been determined by means of X-ray powder diffraction. In addition, the technique provides information on the phase purity of the considered sample, and can be used to optimize the preparation conditions for the synthesis of the cuprate superconductor in question. It therefore seems useful not only to describe the crystal structure, but also to show typical X-ray powder diffraction patterns. The X-ray powder diffraction pattern of $La_{2-x}Sr_xCuO_4$ ($x = 0.15$) is shown. The sketch is based on measurements using Ni-filtered Cu K radiation. All peaks can be assigned to the La-214 phase, and most of them are labeled with the corresponding Miller indices (Chaillout et al., 1989; Hord et al., 2011).

The lattice parameters are a = 0.3779nm and c = 1.3200nm for x = 0.15. Characteristics features of La_2CuO_4 are the CuO_2 planes at the top, the bottom, and in the center of the unit cell (Akhtar et al., 1988; Coffey et al., 1990; Radaelli et al., 1993) (Figure 6.6).

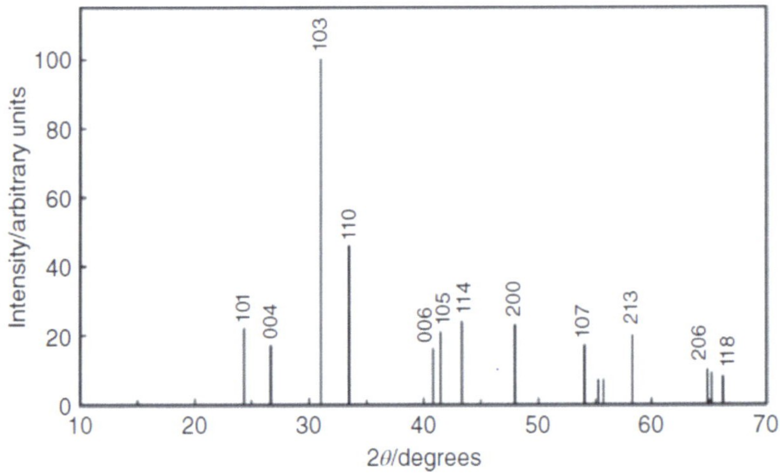

Figure 6.6. *Sketch of the typical Cu Kα X-ray diffraction pattern of $La_{2-x}Sr_{x-}$ CuO_4. For most of the peaks, the Miller indices are indicated.*

Source: Chegg, Creative Commons License.

This square coordinated CuO_2 layers are a common feature of all cuprate high-T_c superconductors. The current in the normal as well as in the superconducting state flows along the CuO_2 planes, which are therefore called conduction planes. In La_2CuO_4, the CuO_2 sheets are separated by insulating double layers of LaO. As shown for the copper ion in the center of the unit cell, each of the copper ions is in the center of an extremely elongated oxygen octahedron. However, the results of band structure calculations indicate that the copper–oxygen interactions along the c axis are small. The unit cell contains two formula units (Das & Saha-Dasgupta, 2009; Tsukada et al.,

2002; Adiperdana et al., 2014). The crystal structure of La_2CuO_4 belongs to the space group *I4/mmm*. Omitting the positions of equivalent atoms, and taking into account the symmetry properties, it is sufficient to provide the normalized positions of only four atoms, there exists an equivalent site at the position $z' = 1 - z$. From the symmetry operation, called body centering, additional atom positions at $x' = x \pm \frac{1}{2}$, $y' = y \pm \frac{1}{2}$; $z' = z \pm \frac{1}{2}$ are resulted. Applying this symmetry operation exchanges edge and centered atoms. Face atoms are moved to another face site (Longo & Raccah, 1973; Howard et al., 1989; Reehuis et al., 2006).

6.3.2. The Crystal Structure of $YBa_2Cu_3O_{7-\delta}$

Next, let us consider the crystal structure of $YBa_2Cu_3O_{7-}\delta$, the currently most important high-temperature superconductor for magnet and power applications. A sketch of the typical X-ray powder diffraction. The indicated Miller indices are based on an orthorhombic unit cell with the lattice parameters $a = 0.38218$nm, $b = 0.38913$nm, and $c = 1.1677$nm (Beno et al., 1987; Lee et al., 2003) (Table 6.1).

Table 6.1. Tabular Representation of Normalized Atom Positions in the Tetragonal Unit Cell of $La_{1.85}Sr_{0.15}CuO_4$ (the Lattice Constants are a = 0.3779 mm and c = 1.3200 nm)

Atoms	*x*	*y*	*z*	**At-oms**	*x*	*y*	*z*
La or Sr	0	0	–0.361	O	0	–0.5	–0.5
La or Sr	–0.5	0.5	–0.139	O	–0.5	0	–0.5
La or Sr	0.5	0.5	–0.139	O	0	0.5	–0.5
La or Sr	0.5	–0.5	–0.139	O	0.5	0	–0.5
La or Sr	–0.5	–0.5	–0.139	O	–0.5	0.5	–0.316
La or Sr	–0.5	0.5	0.139	O	0.5	0.5	–0.316
La or Sr	0.5	0.5	0.139	O	0.5	–0.5	–0.316
La or Sr	0.5	–0.5	0.139	O	–0.5	–0.5	–0.316
La or Sr	–0.5	–0.5	0.139	O	0	0	–0.184
La or Sr	0	0	0.361	O	0	–0.5	0
Cu	–0.5	0.5	–0.5	O	–0.5	0	0
Cu	0.5	0.5	–0.5	O	0	0.5	0
Cu	0.5	–0.5	–0.5	O	–0.5	0	0

Cu	–0.5	–0.5	–0.5	O	0	0	0.184
Cu	0	0	0	O	–0.5	0.5	0.316
Cu	–0.5	0.5	0.5	O	0.5	0.5	0.316
Cu	0.5	0.5	0.5	O	0.5	–0.5	0.316
Cu	0.5	–0.5	0.5	O	–0.5	–0.5	0.316
Cu	–0.5	–0.5	0.5	O	0	–0.5	0.5
–	–	–	–	O	–0.5	0	0.5
–	–	–	–	O	0	0.5	0.5
–	–	–	–	O	0.5	0	0.5

Source: Barbour, Creative Commons License.

Table 6.2. Tabular Representation of Minimum Set of Normalized Atom Positions Required to Describe the Unit Cell of $La_{2-x}Sr_xCuO_4$

Atoms	*x*	*y*	*z*	**Comments**
La or Sr	0	0	0.361	Edge atoms
Cu	0	0	0	Edge atoms
O1	0	0.5	0	Face-atoms
O2	0	0	0184	Edge atoms

Source: Barbour, Creative Commons License.

A space group *Pmmm* is consistent with both X-ray and neutron diffraction data. The crystal structure of the closely related non-superconducting phase $YBa_2Cu_3O_6$ can be described by a tetragonal unit cell (space group *P4/mmm*) of nearly identical cell dimensions (Günther et al., 1999; Schilling et al., 1996).

The unit cell contains the atoms corresponding to one formula unit. The presence of CuO chains along the crystallographic *b* direction is a special feature of the crystal structure of $YBa_2Cu_3O_{7-}\delta$. Each copper atom in the CuO chains is surrounded by four oxygen atoms. (Roeser et al., 2008; Zhu et al., 1991).

As in other cuprate superconductors, CuO_2 planes are present in the crystal structure. Two CuO_2 planes, separated by a layer of Y atoms, form the CuO_2 blocks of Y-123. The slightly different *z* positions of the copper and oxygen atoms lead to a buckling of the CuO_2 planes. Each of the copper atoms in the CuO_2 layers is in the center of the basis of a pyramid formed by five oxygen atoms. Adjacent CuO_2 blocks are separated by double layers of BaO with embedded CuO chains (Pan et al., 1992; Dadras et al., 2009). The Cu1 atoms together with the O_4 atoms form the CuO chains. The lattice sites of the Cu_2, O_2, and O_3 atoms are all in the slightly buckled CuO_2 planes. The oxygen atoms occupy perovskite-like anion positions halfway between the copper ions. The O1 atoms are in the BaO layers, with no oxygen atoms present in the Y layers (Mellekh et al., 2007; Guner et al., 2012).

Frequently, the formula $Ba_2YCu_3O_{7-}\delta$ is used instead of $YBa_2Cu_3O_{7-x}$. This alternative formula takes into account that the Y layers separate the two CuO_2 layers in the CuO_2 blocks, like the Ca layers in the Bi-, Tl-, and Hg-based cuprate superconductors (Shaw et al., 1990; Maeno et al., 1987).

6.3.3. The Crystal Structures of Bi-22(n–1)$_n$ High-Temperature Superconductors

In this section, we will consider the crystal structures of the bismuth-based cuprate high temperature superconductors. There exists a series of compounds described by the chemical formula $Bi_2Sr_2Ca_{n-1}Cu_nO_{2n+4}$. In single samples, the 2201, 2212, and 2223 phases may coexist. It is therefore of importance to study the phase purity by X-ray powder diffraction. In case of the Bi-2223 compound, it proved to be very difficult to synthesize single-phase material without partial substitution of bismuth by lead. We will discuss here only the crystal structures of Bi-2212 and Bi-2223, the two most important bismuth-based cuprate superconductors. These two compounds are used in the first generation of high-T_c superconducting wires and tapes (Wesche, 2017; Bozovic, 2001).

The X-ray powder diffraction pattern of Bi-2212. The Miller indices provided in the figure are based on a pseudotetragonal unit cell with the lattice parameters $a = b = 0.3830$ nm and $c = 3.071$ nm. The true symmetry of Bi-2212 is *Fmmm* or lower. The dimensions of the orthorhombic unit cell based on the space group *Fmmm* are $a = 0.5414$, $b = 0.5418$, and $c = 3.089$ nm. As a special feature, an incommensurate superlattice structure with a period of 4.76 has been observed along the crystallographic *b* direction (Zeljkovic et al., 2014; Verma, 1992).

The Miller indices are based on a tetragonal unit cell with the lattice parameters $a = 0.5411$ nm and $c = 3.707$ nm. In principle, it would have been possible to use a pseudotetragonal unit cell with a lattice parameter ter. It should be noted that, due to the use of different unit cells, the Miller indices provided $\sqrt{a'} = (1/2)a = 0.3826$ nm, as in the case of Bi-2212, and an unchanged *c* lattice parameter (Rodrigues et al., 2017; Owens & Poole,

2002)). For example, the Miller indices (1 1 3) ((Bi,Pb)-2223) and (1 0 3) (Bi-2212) correspond to analogous lattice planes (Haller & Snyder, 1997; Kubiak et al., 1990).

The CuO_2 blocks of Bi-2212 are formed of two CuO_2 sheets, separated by a layer of Ca ions. Neighboring CuO_2 blocks are sandwiched between charge carrier reservoirs with the layer sequence SrO–BiO–BiO–SrO. The copper atoms in the CuO_2 sheets are at the center of the basal plane of a pyramid formed of five oxygen atoms. The Bi-2223 structure contains three CuO_2 sheets in a single CuO_2 block, which are separated by layers of Ca ions. In the central CuO_2 sheet, each copper atom is surrounded by four oxygen atoms. The copper atoms in the outer CuO_2 sheets are in the center of the basal plane of an oxygen pyramid. Adjacent CuO_2 blocks are separated by multilayers of SrO–BiO–BiO–SrO, as in the case of Bi-2212. The insertion of an additional CuO_2–Ca unit transforms the structure of Bi-2212 to that of Bi-2223 (Mohsin & Al-Shakarchi, 2021; Wesche, 2013).

The positions are based on pseudotetragonal unit cells (*I4/mmm*). The application of the body-centering operation provides additional atom position S. In the unit cells of Bi-2212 and Bi-2223, the position $z=0$ is in the central plane of the unit cell (Plakida & Plakida, 2010; Wesche & Wesche,1998).

6.3.4. The Crystal Structures of Tl-Based High-Temperature Superconductors

There exist two families of Tl-based cuprate high-temperature superconductors. The chemical formula of the members of the first family is $Tl_2Ba_2Ca_{n-1}Cu_nO_{2n+4}$ (Tl-22(n–1)$_n$). The Tl-22(n–1)$_n$ compounds are the counterparts of the Bi-22(n–1)$_n$ cuprate superconductors. The second family of Tl-based cuprate high-temperature superconductors is described by the chemical formula $TlM_2Ca_{n-1}Cu_nO_{n+3}$ (Tl-12(n–1)$_n$), where the M atoms are Sr or Ba. As in the Bi-based superconductors, different Tl-based cuprate superconductors may coexist in a single sample. X-ray powder diffraction is therefore a useful tool to ensure that the sample under investigation contains only a single phase. Let us first consider the X-ray powder diffraction patterns and the crystal structures of Tl-22(n – 1)n high-T_c superconductors (de Leon et al., 1994; Civale, 1997).

All reflections can be indexed to an orthorhombic unit cell of space group *Fmmm* with the lattice parameters $a = 0.5473$ nm, $b = 0.5483$ nm, and $c = 2.328$ nm (Pickett, 1989; Moreland et al., 1989).

In principle, the data are also consistent with a tetragonal crystal structure (space group *I4/mmm*). The tetragonal unit cell of $Tl_2Ba_2CuO_6$ (Tl-2201) is shown. Each of the conducting CuO_2 blocks of Tl-2201 contains only a single layer of CuO_2. Neighboring CuO_2 blocks are separated by BaO–TlO–TlO–BaO layers, which form the charge carrier reservoirs of insulating character. Each copper atom is surrounded by six oxygen atoms, forming an octahedron. It should be noted that, as in the case of the Bi-22(n–1)$_n$ high-T_c superconductors, additional atom positions are provided by body centering. The $z = 0$ plane is at the bottom of the tetragonal unit cell (Bellingeri et al., 1999; Morris et al., 1989).

A sketch of a typical X-ray powder diffraction pattern of Tl-2212 made with Cu K radiation. The Miller indices are based on a pseudotetragonal unit cell ($a = b = 0.544$ nm, $c = 2.955$ nm) (De Leon et al., 1993; Apostol et al., 1990).

Note that the unit cell of Tl-2212, the BiO double layers of their Bi-2212 counterparts are replaced by√ for the selected unit cell the lattice parameter a is only (1/2) × 0.544nm = 0.385nm. In TlO double layers, while the SrO layers present in Bi-2212 are replaced by BaO layers. As in their Bi-2212 counterpart, the copper oxide blocks consist of two CuO_2 sheets, separated by a layer of Ca ions. Each Cu atom in the CuO_2 planes is surrounded by a pyramid of five oxygen atoms. (Li et al., 1994; Wijngaarden et al., 1999).

The Miller indices are based on a pseudotetragonal unit cell with the lattice parameters $a = b = 0.540$nm, $c = 3.625$nm. The measurements have been performed using Co Kα radiation ($\lambda = 0.1798$nm). The Miller indices are based on a tetragonal bodycentered unit cell (space group *I4/mmm*) with the lattice parameters $a = 0.3851$nm and $c = 4.201$nm (Dahiya & Rani, 2022; Pattanaik et al., 2012) (Table 6.3).

Table 6.3. Tabular Representation of Normalized Atom Positions in the Tetragonal Unit Cell (I4/mmm) of Tl-2212 (the Lattice Parameters are a ≅ 0.39 nm and c ≅ 2.94 nm)

Atoms	*x*	*y*	*Z*	**Comments**
Tl	0.5	0.5	0.21	Centered atoms[a]
Ba	0	0	0.12	Edge atoms[a]
Cu	0.5	0.5	0.05	Centered atoms[a]
Ca	0	0	0	Edge atoms[a]
O_1	0	0.5	0.05	Face atoms

O_2	0.5	0.5	0.15	Centered atoms[a]
O_3	0	0	0.22	Edge atoms[a]

[a]Atoms at centered (edge) positions in z planes move to edge (centered) positions in z′=z ± 0.5 planes.

Source: R. M. Hazen, Creative Commons License.

The tetragonal unit cells of Tl-2223 ($a = b = 0.382$nm, $c = 3.625$nm) and Tl-2234 ($a = b = 0.385$nm and $c = 4.201$nm) are compared. The sequence of the layers in the charge carrier reservoirs of Tl-2223 and Tl-2234 is BaO–TlO–TlO–BaO, which is identical in all Tl-22(n–1)$_n$ compounds. However, the number n of CuO_2 sheets per copper oxide block increases from 2 in the Tl-2212 to 3 in Tl-2223≥ and 4 in Tl-2234. In all Tl-22(n–1)$_n$ compounds with n 2, the CuO_2 sheets in an individual copper oxide block are separated by a layer of Ca ions. In principle, the insertion of an additional (CuO_2–Ca) block leads from Tl-22(n–1)$_n$ to the next member Tl-22n(n+1) of the Tl cuprate family (Speller, 2003; Peets et al., 2010).

For both compounds, the atom positions are based on a body-centered tetragonal unit cell (*I4*/*mmm*) (Saxena & Saxena, 2012; Cava, 1989).

As an example of the second family of thallium-based cuprate superconductors, we consider $TlBa_2Ca_2Cu_3O_9$. A sketch of the X-ray powder diffraction pattern of Tl-1223. The crystal structure of Tl-1223 belongs to the space group *P4*/*mmm*.

Note that in the primitive tetragonal unit cell, there is no exchange of edge and center atoms by changing the z position from z to $z' = z \pm 0.5$. As in the Tl-2223 counterpart, the copper oxide blocks consist of three CuO_2 sheets. Neighboring CuO_2 planes are separated by a layer of calcium ions. In the central CuO_2 sheet, each of the copper atoms is surrounded by four oxygen atoms, while in the outer CuO_2 planes, the copper atoms are in the center of the basal plane of an oxygen tetrahedron.

A common feature of the Tl-12(n–1)$_n$ compounds is a charge carrier reservoir containing a single TlO layer. The sequence of the layers in the charge carrier reservoirs is BaO–TlO–BaO. The Tl-1212 compound contains one (CuO_2–Ca) unit less than Tl-1223. (Ji et al., 2019; Barbour et al., 1992) (Table 6.4).

Table 6.4. Tabular Representation of Normalized Atom Positions in the Tetragonal Unit Cell (I4/mmm) of Tl-2234 (the Lattice Parameters are a = 0.3850 nm and c = 4.1984 nm)

Atoms	*x*	*y*	*z*	**Comments**
Tl	0.5	0.5	0.226	Centered atoms[a]
Ba	0	0	0.16	Edge atoms[a]
Cu_1	0.5	0.5	0.038	Centered atoms[a]
Cu_2	0.5	0.5	0.113	Edge atoms[a]
Ca_1	0	0	0	Edge atoms[a]
Ca_2	0	0	0.077	Edge atoms[a]
O_1	0.5	0	0.038	Face atoms
O_2	0.5	0	0.113	Face atoms
O_3	0.5	0.5	0.178	Centered atoms[a]
O_4	0.5	0.5	0.268	Centered atoms[a]

Source: D.M. Ogborne & M. T. Weller, Creative Commons License.

6.3.5. The Crystal Structures of Hg-Based High-Temperature Superconductors

The mercury-based cuprates $HgBa_2Ca_{n-1}Cu_nO_{2n+2}$ are structurally closely related to the Tl-12(*n*–1)$_n$ compounds. They are distinguished by the replacement of the TlO layers by HgO layers. The Miller indices are based on a primitive tetragonal unit cell with the lattice parameters *a*=0.388nm and *c*=0.953nm. Hg-1201 crystallizes with the symmetry of the space group *P4*/*mmm* (Wang & Hermann, 1995; Wu et al., 2010).

In the crystal structure of $HgBa_2CuO_4$, single CuO_2 layers are sandwiched between charge carrier reservoirs formed of a BaO–HgO–BaO block. Only 6.3% of the oxygen sites in the HgO layers are occupied. The copper atoms sit in the center of an oxygen octahedron (Barišić et al., 2008; Wu et al., 1999).

All compounds of the mercury-based high-T_c superconductor family Hg-12(*n*–1)$_n$ crystallize with the symmetry of the space group *P4*/*mmm*. In this cuprate family, compounds with more than five CuO_2 sheets per copper oxide block have been successfully synthesized (Roeser et al., 2011; Aslan et al., 2009) (Figures 6.7 and 6.8).

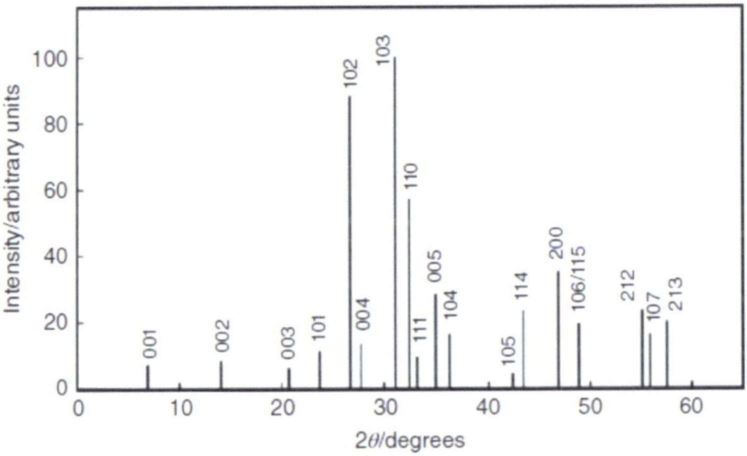

Figure 6.7. *Sketch of the X-ray powder diffraction pattern of Hg-1212 (Cu Kα radiation). The Miller indices are based on a primitive tetragonal unit cell with the lattice parameters a = 0.3862 nm and c = 1.2707 nm.*

Source: N. Barišić, Creative Commons License.

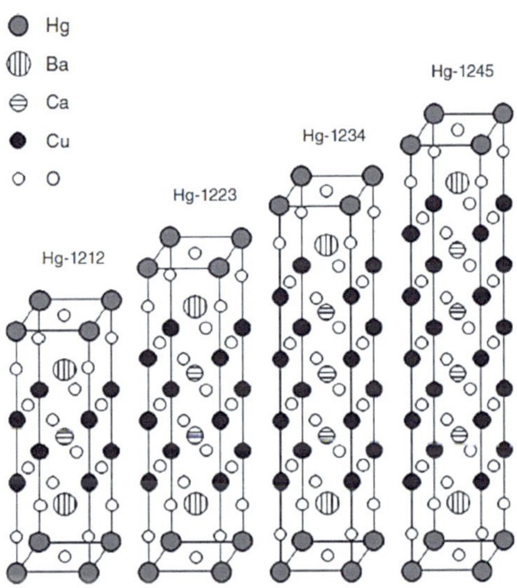

Figure 6.8. *Representation of comparison of the primitive tetragonal unit cells of the Hg-12(n–1)$_n$ cuprate superconductors with n = 2, 3, 4, and 5.*

Source: N. Barišić, Creative Commons License.

In each of these compounds, neighboring CuO_2 sheets within a single copper oxide block are separated by layers of Ca ions. The sequence of the layers in the charge carrier reservoirs is BaO–HgO–BaO, as in the single-layer compound Hg-1201 (He et al., 2002; Tendeloo et al., 1999).

Did You Know?

Lattice parameters, which define the structural dimensions of cuprate superconductors, play a crucial role in understanding their electronic and superconducting properties.

6.3.6. Lattice Parameters of Cuprate Superconductors

Most of the cuprate high-T_c superconductors discussed here crystallize with an orthorhombic or tetragonal crystal structure. In the former case, the difference between the lattice parameters *a* and *b* is very small. Because of the common structural feature of the CuO_2 planes, the lattice parameters *a* and *b* are both close to 0.38nm. In most of the cuprate high-T_c superconductor families, the number of CuO_2 sheets within the CuO_2 blocks can be varied from 1 to *n*, leading to a systematic increase of the *c* lattice parameter with increasing *n*. In general, the lattice parameters of cuprate superconductors are intimately connected with the exact chemical composition. The lattice parameters of a particular cuprate superconductor therefore vary depending on the different synthesis conditions (Nunner et al., 2005; Sakakibara et al., 2010) (Tables 6.5–6.8).

Table 6.5. Tabular Representation of Normalized Atom Positions in the Primitive Tetragonal Unit Cell (P4/mmm) of Hg-1212 (the Lattice Parameters are a = b = 0.385 nm and c = 1.275 nm)

Atoms	*x*	*y*	*z*	**Comments**
Hg	0	0	0	Edge atoms
Ba	0.5	0.5	0.22	Centered atoms
Cu	0	0	0.375	Edge atoms
Ca	0.5	0.5	0.5	Centered atom
O_1	0	0.5	0.38	Face atoms
O_2	0	0	0.154	Edge atoms
O_3^a	0.5	0.5	0	Centered atoms

Source: M. Cantoni, Creative Commons License.

Table 6.6. Tabular Representation of Normalized Atom Positions in the Primitive Tetragonal Unit Cell (P4/mmm) of Hg-1223 (the Lattice Parameters are a = b = 0.385 nm and c = 1.595 nm)

Atoms	*x*	*y*	*z*	**Comments**
Hg	0	0	0	Edge atoms
Ba	0.5	0.5	0.18	Centered atoms
Cu_1	0	0	0.5	Edge atoms
Cu_2	0	0	0.302	Edge atoms
Ca	0.5	0.5	0.4	Centered atom
O_1	0	0.5	0.5	Face atoms
O_2	0	0.5	0.304	Face atoms
O_3	0.5	0.5	0	Centered atoms
O_4[a]	0	0	0.124	Edge atoms

Note: Only 6% of O_4 sites are occupied.

Source: M. Cantoni, Creative Commons License.

Table 6.7. Tabular Representation of Normalized Atom Positions in the Primitive Tetragonal Unit Cell (P4/mmm) of Hg-1234 (the Lattice Parameters are a = 0.385 nm and c = 1.894 nm)

Atoms	*x*	*y*	*z*	**Comments**
Hg[a]	0	0	0	Edge atoms
Ba	0.5	0.5	0.144	Centered atoms
Cu_1	0	0	0.248	Edge atoms
Cu_2	0	0	0.415	Edge atoms
Ca_1	0.5	0.5	0.329	Centered atoms
Ca_2	0.5	0.5	0.5	Centered atoms
O_1	0	0.5	0.251	Face atoms
O_2	0	0.5	0.414	Face atoms
O_3	0	0	0.091	Edge atoms
O_4[b]	0.5	0.5	0	Centered atoms

Note: Only 79% of the Hg sites are occupied. Only 60% of O4 sites are occupied.

Source: S.M. Loureiro et al.; Creative Commons License.

Table 6.8. Tabular Representation of Lattice Parameters of Various Cuprate High-T_c Superconductors

Compound	Symmetry	*a* (nm)	*b* (nm)	*c* (nm)	T_c (K)
$La_{1.9}Sr_{0.1}CuO_4$	Tetragonal	0.3784	0.3784	1.3211	33 [36]
$La_{1.875}Sr_{0.125}CuO_4$	Tetragonal	0.3778	0.3778	1.3216	36 [36]
$La_{1.85}Sr_{0.15}CuO_4$	Tetragonal	0.3779	0.3779	1.3200	36 [6]
$La_{1.85}Sr_{0.15}CuO_4$	Tetragonal	0.3779	0.3779	1.323	39 [36]
$La_{1.775}Sr_{0.225}CuO_4$	Tetragonal	0.3771	0.3771	1.3247	29 [36]
$La_{1.725}Sr_{0.275}CuO_4$	Tetragonal	0.3767	0.3767	1.3225	22 [36]
$Nd_{1.85}Ce_{0.15}CuO_{3.93}$	Tetragonal	0.395	0.395	1.207	24 [12]
$YBa_2Cu_3O_{6.67}$	Orthorhombic	0.3831	0.3889	1.1736	60 [36]
$YBa_2Cu_3O_{6.9}$	Orthorhombic	0.3822	0.3891	1.1677	91 [8]
$YBa_2(Cu_{0.975}Zn_{0.025})_3O_7$	Orthorhombic	0.3820	0.3890	1.1673	62 [36]
$YBa_2(Cu_{0.95}Zn_{0.05})_3O_7$	Orthorhombic	0.3820	0.3885	1.1671	36 [36]
$HoBa_2Cu_3O_7$	Orthorhombic	0.3846	0.3881	1.1640	92 [36]
$YBa_2Cu_4O_8$	Orthorhombic	0.3839	0.3869	2.7243	80 [12]
$HoBa_2Cu_4O_8$	Orthorhombic	0.3855	0.3874	2.7295	80 [36]
$Bi_2Sr_2CuO_6+\delta$	Orthorhombic	0.5361	0.5370	2.4369	8.5 [36]
$Bi_2(Sr_{1.6}La_{0.4})CuO_6+\delta$	Orthorhombic	0.5370	0.5400	2.450	23.5 [36]
$Bi_{2.2}Sr_2Ca_{0.8}Cu_2O_8+\delta$	Orthorhombic	0.5414	0.5418	3.089	84 [17]

$Bi_2Sr_2CaCu_2O_8$	Orthorhombic	0.5413	0.5411	3.091	89 [36]
$Bi_2Sr_2CaCu_2O_8+\delta$	Orthorhombic	0.5408	0.5413	3.081	75 [36]
$Bi_2Sr_2CaCu_2O_8+\delta$	Orthorhombic	0.5407	0.5415	3.0874	83 [37]
$Bi_2Sr_{1.5}Ca_{1.5}Cu_2O_{8.19}$	Orthorhombic	0.5402	0.5408	3.0699	83 [37]
$Bi_2Sr_{1.5}Ca_{1.5}Cu_2O_{8.22}$	Orthorhombic	0.5396	0.5400	3.0635	73 [37]
$Bi_2Sr_{1.5}Ca_{1.5}Cu_2O_{8.22}$	Orthorhombic	0.5396	0.5404	3.0618	67 [37]
$Bi_2Sr_2Ca_{0.9}Y_{0.1}Cu_2O_{8.23}$	Orthorhombic	0.5408	0.5416	3.0807	90 [37]
$Bi_2Sr_2Ca_{0.7}Y_{0.3}Cu_2O_{8.30}$	Orthorhombic	0.5410	0.5425	3.0619	86 [37]
$Bi_2Sr_2Ca_2Cu_3O_{10}+\delta$	Orthorhombic	0.541	0.539	3.82	115 [18]
$Bi_{1.7}Pb_{0.25}Sr_{1.47}Ca_{2.54}Cu_{3.1}O_{10}+\delta$	Orthorhombic	0.5411	0.5411	3.707	105 [16]
$(Bi,Pb)_2Sr_{1.72}Ca_2Cu_3O_{10}+\delta$	Orthorhombic	0.5392	0.5395	3.6985	111 [38]
$(Bi_{1.6}Pb_{0.4})Sr_2Ca_2Cu_3O_{10}$	Orthorhombic	0.5413	0.5413	3.7100	107 [36]
$Tl_2Ba_2CuO_6$	Tetragonal	0.3866	0.3866	2.3239	85 [36]
$Tl_2Ba_2CuO_6+\delta$	Orthorhombic	0.5473	0.5483	2.3277	93 [19]
$Tl_{1.7}Ba_2Ca_{1.06}Cu_{2.32}O_8+\delta$	Tetragonal	0.3857	0.3857	2.939	108 [12]
$Tl_2Ba_2CaCu_2O_8$	Tetragonal	0.3855	0.3855	2.9318	99 [36]
$Tl_2Ba_2Ca_2Cu_3O_{10}$	Tetragonal	0.3850	0.3850	3.588	125 [36]
$Tl_{1.64}Ba_2Ca_{1.87}Cu_{3.11}O_{10}+\delta$	Tetragonal	0.3822	0.3822	3.626	125 [12]
$Tl_2Ba_2Ca_3Cu_4O_{12}+\delta$	Tetragonal	0.3850	0.3850	4.1984	114 [24]
$Tl_{1.1}Ba_2Ca_{0.9}Cu_{2.1}O_{7.1}$	Tetragonal	0.3851	0.3851	1.2728	80 [39]

$Tl_{1.1}Ba_2Ca_{1.8}Cu_{3.0}O_{9.7}$	Tetragonal	0.3843	0.3843	1.5871	110 [39]
$TlBa_2Ca_3Cu_4O_{12}+\delta$	Tetragonal	0.3848	0.3848	1.9001	114 [40]
$(Tl,Cr)Sr_2(Ca,Tl)Cu_2O_7$	Tetragonal	0.3816	0.3816	1.2022	100 [41]
$(Tl_{0.5}Pb_{0.5})Sr_2CaCu_2O_7$	Tetragonal	0.3802	0.3802	1.2107	80 [36]
$(Tl_{0.5}Pb_{0.5})Sr_2Ca_2Cu_3O_9$	Tetragonal	0.3821	0.3821	1.5294	122 [36]
$HgBa_2CuO_4+\delta$	Tetragonal	0.380	0.380	0.9509	94 [42]
$HgBa_2CaCu_2O_6+\delta$	Tetragonal	0.3859	0.3859	1.2657	123 [43]
$HgBa_2CaCu_2O_6+\delta$	Tetragonal	0.3856	0.3856	1.2652	121 [35]
$HgBa_2Ca_2Cu_3O_8+\delta$	Tetragonal	0.3853	0.3853	1.5818	133 [35]
$HgBa_2Ca_2Cu_3O_8+\delta$	Tetragonal	0.3855	0.3855	1.5842	110 [44]
$HgBa_2Ca_3Cu_4O_{10}+\delta$	Tetragonal	0.3854	0.3854	1.9006	126 [35]
$HgBa_2Ca_4Cu_5O_{12}+\delta$	Tetragonal	0.3852	0.3852	2.2141	110 [45]
$HgBa_2Ca_5Cu_6O_{14}+\delta$	Tetragonal	0.3852	0.3852	2.526	107 [45]
$(Hg_{0.9}Pb_{0.1})Ba_2Ca_2Cu_3O_8+\delta$	Tetragonal	0.3850	0.3850	1.5836	118 [44]
$(Hg_{0.8}Pb_{0.2})Ba_2Ca_2Cu_3O_8+\delta$	Tetragonal	0.3848	0.3848	1.5831	125 [44]
$(Hg_{0.7}Pb_{0.3})Ba_2Ca_2Cu_3O_8+\delta$	Tetragonal	0.3843	0.3843	1.5824	130 [44]
$(Hg_{0.6}Pb_{0.4})Ba_2Ca_2Cu_3O_8+\delta$	Tetragonal	0.3842	0.3842	1.5818	120 [44]
$(Hg_{0.5}Pb_{0.5})Ba_2Ca_2Cu_3O_8+\delta$	Tetragonal	0.3840	0.3840	1.5822	103 [44]
$(Hg_{0.5}Pb_{0.5})Ba_2Ca_4Cu_5O_{12}+\delta$	Tetragonal	0.3853	0.3853	2.2172	115 [12]
$(Tl_{0.5}Pb_{0.5})Sr_2Ca_2Cu_3O_9$	Tetragonal	0.3815	0.3815	1.5280	118 [12]

$BSr_2Ca_2Cu_3O_9+\delta$	Tetragonal	0.3821	0.3821	1.3854	75 [12]
$BSr_2Ca_3Cu_4O_{11}+\delta$	Tetragonal	0.3836	0.3836	1.7082	110 [12]
$BSr_2Ca_4Cu_5O_{13}+\delta$	Tetragonal	0.3837	0.3837	2.022	85 [12]
$AlSr_2Ca_{1.5}Y_{0.5}Cu_3O_9+\delta$	Tetragonal	0.3836	0.3836	1.4405	78 [12]
$AlSr_2Ca_3Cu_4O_{11}+\delta$	Tetragonal	0.3839	0.3839	1.772	110 [12]
$AlSr_2Ca_4Cu_5O_{13}+\delta$	Tetragonal	0.3845	0.3845	2.087	83 [12]

Source: Decroux et al.; Creative Commons License.

The c lattice parameters of the different members of the Tl-22$(n–1)_n$ family are approximately given by the formula $c \cong (1.7 + 0.62n)$nm. The crystal structures of all Tl-22$(n–1)_n$ cuprate high-T_c superconductors are close to an ideal tetragonal structure (symmetry group *I4/mmm*). The structures of the Tl-22$(n–1)_n$ compounds are closely related to those of their Bi-22$(n–1)_n$ counterparts. However, the structural modulations generally found in the bismuth-based compounds are missing. The crystal structures of Tl-12$(n–1)_n$ compounds are primitive tetragonal (space group *P4/mmm*). Their lattice parameters are close to $a \cong 0.38$nm and $c \cong (0.63 + 0.32n)$nm. Their mercury-based counterparts Hg-12$(n–1)_n$ also crystallize with the symmetry of the space group *P4/mmm*. The lattice parameters of the mercury-based compounds are $a \cong 0.39$nm and $c \cong (0.95 + 0.32(n – 1))_{\text{nm}}$ (Butko et al., 2009; Choi et al., 2022).

A further aspect of importance is the systematically different distance of adjacent CuO_2 blocks. In the Bi-22$(n–1)_n$ and Tl-22$(n–1)_n$ compounds, this distance is relatively large as compared to Tl-12$(n–1)_n$ and Hg-12$(n–1)_n$ high-T_c superconductors. For example, the distance between the CuO_2 planes in Tl-2201 is 1.16nm, but only 0.95nm in Hg-1201. The large distance between adjacent CuO_2 blocks in the Bi-22$(n–1)_n$ and Tl-22$(n–1)_n$ compounds seems to cause a weak superconducting coupling of adjacent CuO_2 blocks, which leads to a pronounced anisotropy of the superconducting properties (Tsuei & Kirtley, 2000; Vojta, 2009).

The lattice parameters depend, like many other physical properties, on the exact oxygen content, and hence on the synthesis conditions (Vaughey et al., 1991; Plakida, 2010).

Activity 6.1

You are a crystallographer investigating the crystal structures of cuprate superconductors. Organize a virtual or hands-on activity where participants can build three-dimensional models of cuprate structures and analyze the relationship between their atomic arrangement and superconducting properties.

6.4. SUMMARY

Chapter six focuses on the crystal structures of cuprate superconductors. It begins with an introduction to the chapter's topics and then explores diffraction methods used to analyze crystal structures, including the Bragg condition, Miller indices, classification of crystal structures, X-ray diffraction, and neutron diffraction. The chapter delves into the crystal structures of various cuprate high-temperature superconductors, starting with the crystal structure of La2CuO4. It further discusses the crystal structures of Bi-22$(n-1)_n$ compounds, Tl-based high-temperature superconductors, and Hg-based high-temperature superconductors. The discussion includes the unique characteristics and properties of these crystal structures and their relationship to the superconducting behavior exhibited by cuprate superconductors. Additionally, the chapter covers lattice parameters of cuprate superconductors, highlighting their significance in understanding the electronic and superconducting properties of these materials.

REVIEW QUESTIONS

1. Why is understanding the crystal structures of cuprate superconductors important in studying their high-temperature superconductivity?
2. Explain the diffraction methods used to analyze crystal structures, including X-ray diffraction and neutron diffraction.
3. What is the Bragg condition, and how is it relevant to diffraction methods?
4. Describe the role of Miller indices in classifying crystal structures.
5. Discuss the classification of crystal structures and its significance in understanding cuprate superconductors.
6. How does X-ray diffraction contribute to the analysis of crystal structures in cuprate superconductors?

7. Explain the importance of neutron diffraction in studying the crystal structures of cuprate superconductors.
8. Provide an overview of the crystal structure of La2CuO4 and its relationship to high-temperature superconductivity.

REFERENCES

1. Adiperdana, B., Suprayoga, E., Adam, N., Mohd-Tajudin, S. S., Rozlan, A. F., Sulaiman, S., & Watanabe, I. (2014, December). An effect of the supercell calculation on muon positions and local deformations of crystal structure in La2CuO4. In *Journal of Physics: Conference Series* (Vol. 551, No. 1, p. 012051). IOP Publishing.
2. Akhtar, M. J., Catlow, C. R. A., Clark, S. M., & Temmerman, W. M. (1988). The pressure dependence of the crystal structure of La2CuO4. *Journal of Physics C: Solid State Physics*, *21*(25), L917.
3. Apostol, M., Buzatu, F., & Liu, F. H. (1990). Critical temperature of third generation high-temperature superconductors. *International Journal of Modern Physics B*, *4*(01), 159–177.
4. Aslan, Ö., Özdemir, Z. G., Keskin, S. S., & Onbaşli, Ü. (2009, March). The chaotic points and XRD analysis of Hg-based superconductors. In *Journal of Physics: Conference Series* (Vol. 153, No. 1, p. 012002). IOP Publishing.
5. Axe, J. D., & Crawford, M. K. (1994). Structural instabilities in lanthanum cuprate superconductors. *Journal of Low Temperature Physics*, *95*, 271–284.
6. Barbour, J. C., Venturini, E. L., Ginley, D. S., & Kwak, J. F. (1992). Irradiation effects in high temperature superconductors. *Nuclear Instruments and Methods in Physics Research Section B: Beam Interactions with Materials and Atoms*, *65*(1–4), IN9–538.
7. Barišić, N., Li, Y., Zhao, X., Cho, Y. C., Chabot-Couture, G., Yu, G., & Greven, M. (2008). Demonstrating the model nature of the high-temperature superconductor $HgBa_2CuO_{4+}$ δ. *Physical Review B*, *78*(5), 054518.
8. Bellingeri, E., Grasso, G., Gladyshevskii, R., Giannini, E., Marti, F., Dhalle, M., & Flükiger, R. (1999). Fluorine substitution in high temperature superconductors. *International Journal of Modern Physics B*, *13*(09n10), 973–978.
9. Beno, M. A., Soderholm, L., Capone, D. W., Hinks, D. G., Jorgensen, J. D., Grace, J. D., & Zhang, K. (1987). Structure of the single-phase high-temperature superconductor $YBa2Cu3O_{7-\delta}$. *Applied Physics Letters*, *51*(1), 57–59.
10. Betzel, C., Saenger, W., Hingerty, B. E., & Brown, G. M. (1984). Topography of cyclodextrin inclusion complexes, part 20. Circular and flip-flop hydrogen bonding in. beta.-cyclodextrin undecahydrate:

a neutron diffraction study. *Journal of the American Chemical Society, 106*(24), 7545–7557.

11. Birkl, G., Gatzke, M., Deutsch, I. H., Rolston, S. L., & Phillips, W. D. (1995). Bragg scattering from atoms in optical lattices. *Physical Review Letters, 75*(15), 2823.
12. Bozovic, I. (2001). Atomic-layer engineering of superconducting oxides: yesterday, today, tomorrow. *IEEE Transactions on Applied Superconductivity, 11*(1), 2686–2695.
13. Butko, V. Y., Logvenov, G., Božović, N., Radović, Z., & Božović, I. (2009). Madelung strain in cuprate superconductors – a route to enhancement of the critical temperature. *Advanced Materials, 21*(36), 3644–3648.
14. Cao, Y., & Tan, H. (2005). Study on crystal structures of enzyme-hydrolyzed cellulosic materials by X-ray diffraction. *Enzyme and Microbial Technology, 36*(2–3), 314–317.
15. Cava, R. J. (1989). Crystal Chemistry of the Copper Oxide Based High Temperature Superconductors. In *Advances in Superconductivity: Proceedings of the 1st International Symposium on Superconductivity (ISS'88), August 28–31, 1988, Nagoya* (pp. 159–164). Springer Japan.
16. Chaillout, C., Cheong, S. W., Fisk, Z., Lehmann, M. S., Marezio, M., Morosin, B., & Schirber, J. E. (1989). The crystal structure of superconducting La2CuO4. 032 by neutron diffraction. *Physica C: Superconductivity, 158*(1–2), 183–191.
17. Choi, J., Wang, Q., Jöhr, S., Christensen, N. B., Küspert, J., Bucher, D., & Chang, J. (2022). Unveiling unequivocal charge stripe order in a prototypical cuprate superconductor. *Physical Review Letters, 128*(20), 207002.
18. Civale, L. (1997). Vortex pinning and creep in high-temperature superconductors with columnar defects. *Superconductor Science and Technology, 10*(7A), A11.
19. Coffey, D., Bedell, K. S., & Trugman, S. A. (1990). Effective spin Hamiltonian for the CuO planes in La 2 CuO 4 and metamagnetism. *Physical Review B, 42*(10), 6509.
20. Dadras, S., Liu, Y., Chai, Y. S., Daadmehr, V., & Kim, K. H. (2009). Increase of critical current density with doping carbon nano-tubes in YBa2Cu3$O_{7-\delta}$. *Physica C: Superconductivity, 469*(1), 55–59.
21. Dahiya, S., & Rani, K. (2022). Study of high-temperature superconductivity. *Physical Sciences, 1*, 17.
22. Das, H., & Saha-Dasgupta, T. (2009). Electronic structure of La2CuO4 in

the T and T′ crystal structures using dynamical mean field theory. *Physical Review B*, *79*(13), 134522.

23. De Leon, J. M., Conradson, S. D., Bishop, A. R., & Raistrick, I. D. (1993). The local structure of high-temperature superconductors. *Japanese Journal of Applied Physics*, *32*(S2), 573.
24. de Leon, J. M., Li, G. G., Conradson, S. D., Bishop, A. R., Subramanian, M. A., & Raistrick, I. D. (1994). Planar oxygen-centered lattice instabilities in Tl-based high-temperature superconductors. *Physica C: Superconductivity, 220*(3–4), 377–382.
25. Dong, B. (2018). *Neutron Diffraction Studies of Lithium-based Battery Materials* (Doctoral dissertation, University of Sheffield).
26. Dunlop, R. S. (1973). *Crystal Structure Studies by X-Ray and Neutron Diffraction*. University of Glasgow (United Kingdom).
27. Dyson, P. J., & McIndoe, J. S. (2000). *Transition Metal Carbonyl Cluster Chemistry* (Vol. 2). CRC Press.
28. Guner, S. B., Gorur, O., Celik, S., Dogruer, M., Yildirim, G., Varilci, A., & Terzioglu, C. (2012). Effect of zirconium diffusion on the microstructural and superconducting properties of $YBa2Cu3O_{7-\delta}$ superconductors. *Journal of Alloys and Compounds*, *540*, 260–266.
29. Günther, W., Schöllhorn, R., Epple, M., Siegle, H., Thomsen, C., Kabius, B., & Zander, W. (1999). Hydrogen and water intercalation into $YBa2Cu3O_{7-\delta}$: Structural properties of $H2YBa2Cu3O_{8-\delta}$. *Philosophical Magazine A*, *79*(2), 449–466.
30. Haller, M., & Snyder, R. L. (1997). The structural conditions for high-temperature superconductivity. *JOM*, *49*(10), 12–17.
31. He, H., Bourges, P., Sidis, Y., Ulrich, C., Regnault, L. P., Pailhes, S., & Keimer, B. (2002). Magnetic resonant mode in the single-layer high-temperature superconductor Tl2Ba2CuO6+ δ. *Science*, *295*(5557), 1045–1047.
32. Hord, R., Cordier, G., Hofmann, K., Buckow, A., Pascua, G., Luetkens, H., & Albert, B. (2011). Transitions Between Lanthanum Cuprates: Crystal Structures of T′, Orthorhombic, and K2NiF4-type La2CuO4, 5, 67-90.
33. Howard, C. J., Nelmes, R. J., & Vettier, C. (1989). A neutron powder diffraction study of the effect of pressure on the crystal structure of La2CuO4. *Solid state communications*, *69*(3), 261–264.
34. Ji, L., Liang, X., He, K., Xing, J., Xue, T., Xu, T., & Feng, M. (2019). Growth and property analysis of large-area epitaxial Tl-2212 high temperature superconducting films. *Ceramics International*, *45*(18), 24635–24641.

35. Kubiak, R., Westerholt, K., Pelka, G., Bach, H., & Khan, Y. (1990). Pressure dependence of the superconducting transition temperature of Bi-and Tl-based high-Tc superconductors. *Physica C: Superconductivity, 166*(5–6), 523–529.

36. Kudoh, Y., Tokonami, M., Miyazaki, S., & Otsuka, K. (1985). Crystal structure of the martensite in Ti-49.2 at.% Ni alloy analyzed by the single crystal X-ray diffraction method. *Acta Metallurgica, 33*(11), 2049–2056.

37. Lee, J. H., Ko, Y. B., & Lee, S. Y. (2003). Correlation between the microwave surface resistance and the crystal structures of YBa2Cu3$O_{7-\delta}$ films on CeO2-buffered sapphire. *Superconductor Science and Technology, 16*(3), 386.

38. Li, G. G., de Leon, J. M., Conradson, S. D., Lovato, M. V., & Subramanian, M. A. (1994). Local distortions of the Tl-O layers in Tl-based high-temperature superconductors. *Physical Review B, 50*(5), 3356.

39. Longo, J. M., & Raccah, P. M. (1973). The structure of La2CuO4 and LaSrVO4. *Journal of Solid State Chemistry, 6*(4), 526–531.

40. Maeno, Y., Tomita, T., Kyogoku, M., Awaji, S., Aoki, Y., Hoshino, K., & Fujita, T. (1987). Substitution for copper in a high-T c superconductor YBa2Cu3$O_{7-\delta}$. *Nature, 328*(6130), 512–514.

41. Mellekh, A., Zouaoui, M., Azzouz, F. B., Annabi, M., & Salem, M. B. (2007). Effect of nanometer Al2O3 addition on structural and transport properties of YBa2Cu3$O_{7-\delta}$. *Physica C: Superconductivity and Its Applications, 460*, 426–427.

42. Mohsin, A., & Al-Shakarchi, E. K. (2021). The Simulation on the Phase Formation of Bi-Ba-Ca-Cu-O high Temperature Superconductor. *Journal of Advanced Materials and Processing, 9*(4), 35–42.

43. Moreland, J., Ginley, D. S., Venturini, E. L., & Morosin, B. (1989). Break junction measurement of the tunneling gap of a thallium-based high-temperature superconductor crystal. *Applied Physics Letters, 55*(14), 1463–1465.

44. Morris, D. E., Nickel, J. H., Wei, J. Y., Asmar, N. G., Scott, J. S., Scheven, U. M., & Hazen, R. M. (1989). Eight new high-temperature superconductors with the 1: 2: 4 structure. *Physical Review B, 39*(10), 7347.

45. Nandi, S., Xiao, Y., Qureshi, N., Paramanik, U. B., Jin, W. T., Su, Y., & Brückel, T. (2016). Magnetic structures of the Eu and Cr moments in EuCr 2 As 2: Neutron diffraction study. *Physical Review B, 94*(9), 094411.

46. Nunner, T. S., Andersen, B. M., Melikyan, A., & Hirschfeld, P. J. (2005). Dopant-modulated pair interaction in cuprate superconductors. *Physical Review Letters, 95*(17), 177003.

47. O'connor, B. H., & Valentine, T. M. (1969). A neutron diffraction study of the crystal structure of the C-form of yttrium sesquioxide. *Acta Crystallographica Section B: Structural Crystallography and Crystal Chemistry*, *25*(10), 2140–2144.

48. Owens, F. J., & Poole, C. P. (2002). New High-Temperature Superconductors. *The New Superconductors*, 97–119.

49. Pan, V. M., Svetchnikov, V. L., Solovjov, V. F., Taborov, V. F., Zandbergen, H. W., & Wen, J. G. (1992). YBa2Cu3O7-delta single-crystal microstructure related to transport critical current density. *Superconductor Science and Technology*, *5*(12), 707.

50. Park, C., & Snyder, R. L. (1995). Structures of High-Temperature Cuprate Superconductors. *Journal of the American Ceramic Society*, *78*(12), 3171–3194.

51. Park, W. B., Chung, J., Jung, J., Sohn, K., Singh, S. P., Pyo, M., & Sohn, K. S. (2017). Classification of crystal structure using a convolutional neural network. *IUCrJ*, *4*(4), 486–494.

52. Pattanaik, A., Meher, S. K., & Nayak, P. (2012). Magnetization study of Hg and Tl based high temperature superconductors in mixed state. *Journal of Superconductivity and Novel Magnetism*, *25*, 1867–1872.

53. Patterson, A. L. (1944). Ambiguities in the X-ray analysis of crystal structures. *Physical Review*, *65*(5–6), 195.

54. Peets, D. C., Liang, R., Raudsepp, M., Hardy, W. N., & Bonn, D. A. (2010). Encapsulated single crystal growth and annealing of the high-temperature superconductor Tl-2201. *Journal of Crystal Growth*, *312*(2), 344–350.

55. Pickett, W. E. (1989). Electronic structure of the high-temperature oxide superconductors. *Reviews of Modern Physics*, *61*(2), 433.

56. Plakida, N. (2010). *High-Temperature Cuprate Superconductors: Experiment, Theory, and Applications* (Vol. 166). Springer Science & Business Media.

57. Plakida, N., & Plakida, N. (2010). Crystal Structure. *High-Temperature Cuprate Superconductors: Experiment, Theory, and Applications*, 13–50.

58. Radaelli, P. G., Jorgensen, J. D., Schultz, A. J., Hunter, B. A., Wagner, J. L., Chou, F. C., & Johnston, D. C. (1993). Structure of the superconducting La2 CuO4+ δ phases (δ≊ 0.08, 0.12) prepared by electrochemical oxidation. *Physical Review B*, *48*(1), 499.

59. Reehuis, M., Ulrich, C., Prokeš, K., Gozar, A., Blumberg, G., Komiya, S., & Keimer, B. (2006). Crystal structure and high-field magnetism of la 2 cu o 4. *Physical Review B*, *73*(14), 144513.

60. Rodrigues, V. D., Souza, G. A. D., Carvalho, C. L., & Zadorosny, R. (2017). Effect of la doping on the crystal structure, electric, magnetic and morphologic properties of the BSCCO system. *Materials Research*, *20*, 1406–1413.

61. Roeser, H. P., Haslam, D. T., López, J. S., Stepper, M., Von Schoenermark, M. F., Huber, F. M., & Nikoghosyan, A. S. (2011). Electronic energy levels in high-temperature superconductors. *Journal of Superconductivity and Novel Magnetism*, *24*, 1443–1451.

62. Roeser, H. P., Hetfleisch, F., Huber, F. M., Von Schoenermark, M. F., Stepper, M., Moritz, A., & Nikoghosyan, A. S. (2008). A link between critical transition temperature and the structure of superconducting YBa2Cu3O7-δ. *Acta Astronautica*, *62*(12), 733–736.

63. Sakakibara, H., Usui, H., Kuroki, K., Arita, R., & Aoki, H. (2010). Two-orbital model explains the higher transition temperature of the single-layer Hg-cuprate superconductor compared to that of the La-cuprate superconductor. *Physical Review Letters, 105*(5), 057003.

64. Salmon, P. S., & Lond, P. B. (1992). The structure of Cl-solutions in ethylene glycol studied by neutron diffraction. *Journal of Physics: Condensed Matter*, *4*(23), 5249.

65. Saxena, A. K., & Saxena, A. K. (2012). Crystal Structure of High Temperature Superconductors. *High-Temperature Superconductors*, 43–60.

66. Schilling, A., Fisher, R. A., Phillips, N. E., Welp, U., Dasgupta, D., Kwok, W. K., & Crabtree, G. W. (1996). Calorimetric measurement of the latent heat of vortex-lattice melting in untwined YBa2Cu3$O_{7-\delta}$. *Nature*, *382*(6594), 791–793.

67. Shaw, T. M., Dimos, D., Batson, P. E., Schrott, A. G., Clarke, D. R., & Duncombe, P. R. (1990). Carbon retention in YBa2Cu3$O_{7-\delta}$ and its effect on the superconducting transition. *Journal of Materials Research*, *5*(6), 1176–1184.

68. Speller, S. C. (2003). Thallium based high temperature superconductors for microwave device applications. *Materials Science and Technology*, *19*(3), 269–282.

69. Staron, P. (2017). Stress Analysis by Angle-Dispersive Neutron Diffraction. *Neutrons and Synchrotron Radiation in Engineering Materials Science: From Fundamentals to Applications*, 105–122.

70. Sun, S., Zhang, X., Cui, J., & Liang, S. (2020). Identification of the Miller indices of a crystallographic plane: a tutorial and a comprehensive review on fundamental theory, universal methods based on different case studies

and matters needing attention. *Nanoscale*, *12*(32), 16657–16677.

71. Takita, K., Katoh, H., Akinaga, H., Nishino, M., Ishigaki, T., & Asano, H. (1988). X-ray diffraction study on the crystal structure of Nd1+ xBa2-xCu3O7-δ. *Japanese Journal of Applied Physics*, *27*(1A), L57.
72. Tendeloo, G., Van Tendeloo, G., Antipov, E. V., & Putilin, S. N. (Eds.). (1999). *High-Temperature Superconductors and Novel Inorganic Materials* (Vol. 62). Springer Science & Business Media.
73. Tsuei, C. C., & Kirtley, J. R. (2000). Pairing symmetry in cuprate superconductors. *Reviews of Modern Physics*, *72*(4), 969.
74. Tsukada, A., Greibe, T., & Naito, M. (2002). Phase control of La 2 CuO 4 in thin film synthesis. *Physical Review B*, *66*(18), 184515.
75. Tsukada, A., Krockenberger, Y., Noda, M., Yamamoto, H., Manske, D., Alff, L., & Naito, M. (2005). New class of T′-structure cuprate superconductors. *Solid State Communications*, *133*(7), 427–431.
76. Vaughey, J. T., Thiel, J. P., Hasty, E. F., Groenke, D. A., Stern, C. L., Poeppelmeier, K. R., & Mitchell, A. W. (1991). Synthesis and structure of a new family of cuprate superconductors: LnSr2Cu2GaO7. *Chemistry of Materials*, *3*(5), 935–940.
77. Vaz, W. A. (2014). *X-ray and Neutron Diffraction Studies of some Rare Earth Doped Borate Glasses and Coated Silica Microspheres* (Doctoral dissertation, Goa University).
78. Verma, G. D., Ramakrishna, K., & Srivastava, O. N. (1992). On electron microscopic studies of structural characteristics of the Tl (Bi, Pb)SrCaCuO high-temperature superconductor. *Physica C: Superconductivity, 202*(3–4), 327–334.
79. Vojta, M. (2009). Lattice symmetry breaking in cuprate superconductors: stripes, nematics, and superconductivity. *Advances in Physics*, *58*(6), 699–820.
80. Wang, Y. T., & Hermann, A. M. (1995). Influence of fluorine in magnetic properties of Hg-1223 high-temperature superconductors. *Physica C: Superconductivity, 254*(1–2), 1–6.
81. Wesche, R. (2013). *High-temperature superconductors: materials, properties, and applications* (Vol. 6). Springer Science & Business Media.
82. Wesche, R. (2017). High-temperature superconductors. *Springer Handbook of Electronic and Photonic Materials*, 1–1.
83. Wesche, R., & Wesche, R. (1998). A survey of superconducting cuprates. *High-Temperature Superconductors: Materials, Properties, and Applications*, 79–132.

84. Wijngaarden, R. J., Jover, D. T., & Griessen, R. (1999). Intrinsic and carrier density effects on the pressure dependence of Tc of high-temperature superconductors. *Physica B: Condensed Matter*, *265*(1–4), 128–135.

85. Wu, J. Z., Yan, S. L., & Xie, Y. (1999). Cation exchange: A scheme for synthesis of mercury-based high-temperature superconducting epitaxial thin films. *Applied Physics Letters, 74*(10), 1469–1471.

86. Wu, J., Zhao, H., Bhattacharya, R., & Parans, M. (2010). Recent progress in fabrication, characterization, and application of hg-based oxide superconductors. *High Temperature Superconductors*.

87. Zeljkovic, I., Nieminen, J., Huang, D., Chang, T. R., He, Y., Jeng, H. T., & Hoffman, J. E. (2014). Nanoscale interplay of strain and doping in a high-temperature superconductor. *Nano Letters*, *14*(12), 6749–6753.

88. Zhu, Y., Zhang, H., Wang, H., & Suenaga, M. (1991). Grain boundary in textured $YBa2Cu3O_{7-\delta}$ superconductor. *Journal of Materials Research*, *6*(12), 2507–2518.

89. Ziletti, A., Kumar, D., Scheffler, M., & Ghiringhelli, L. M. (2018). Insightful classification of crystal structures using deep learning. *Nature Communications*, *9*(1), 2775.

CHAPTER

7 Practical Applications of High-Temperature Superconductors

Contents

7.1. INTRODUCTION

The chapter thoroughly explains the diverse practical implementations and technological progressions facilitated by high-temperature superconductors (high-Tc superconductors). The text explores the importance of these materials in propelling tangible advancements across various disciplines.

The introductory section of the chapter highlights the significance of pragmatic implementations of high-temperature superconductors. The statement underscores the transformative impact of materials that demonstrate superconductivity at temperatures that exceed those of traditional superconductors on technological progress. The distinctive characteristics of superconductors, including their absence of electrical resistance and ideal diamagnetism, render them exceedingly sought-after for diverse uses.

The primary objective of this chapter is to examine and deliberate upon the multifarious pragmatic implementations of high-temperature superconductors. The text underscores the principal domains these materials have rendered noteworthy contributions to. The chapter encompasses a diverse range of domains, such as superconducting cables and wires, magnetic levitation and transport systems, superconducting magnets, storage of energy, superconducting electronics, and quantum computing, as well as sensing and recognition applications, and emerging applications (Shen et al., 2020).

Every segment contained in the chapter offers a comprehensive examination of the particular field of application. The text elucidates the merits and benefits of utilizing high-temperature superconductors in various domains and deliberates on the role of these materials in facilitating technological progress. The chapter elucidates particular instances and empirical analyses to exemplify high-temperature superconductors' pragmatic applications and tangible ramifications. The chapter also delves into the obstacles and constraints linked to the pragmatic utilization of high-temperature superconductors. The text delves into the various obstacles that must be overcome to achieve successful practical implementation, including but not limited to materials obstacles, pricing considerations, scalability issues, as well as obstacles with existing infrastructure (Hull, 2003).

The primary objective of the chapter is to furnish a thorough exposition of the pragmatic uses of high-temperature superconductors, accentuating their significance in diverse sectors and their prospects for further progress. This highlights the significant influence of these resources in propelling technological progress and molding the prospects of various domains (Weiss et al., 2016).

7.2. SIGNIFICANCE OF HIGH-TEMPERATURE SUPERCONDUCTORS IN TECHNOLOGICAL ADVANCEMENTS

The significance of high-temperature superconductors (HTS) has been on the rise, as they are instrumental in propelling technological progress in diverse domains. Their significance is attributed to their capacity to conduct electrical currents with negligible resistance, resulting in minimal power dissipation. The exceptional characteristics in question has facilitated the advancement of superconductive wires and cables, which can transmit electrical energy with significantly greater efficacy than customary substances. High-temperature superconducting (HTS) systems can transform electrical grids and augment energy efficiency globally by mitigating energy losses during extended power transmission. This can pave the way for a more sustainable and environmentally friendly future (Muzzi et al., 2015).

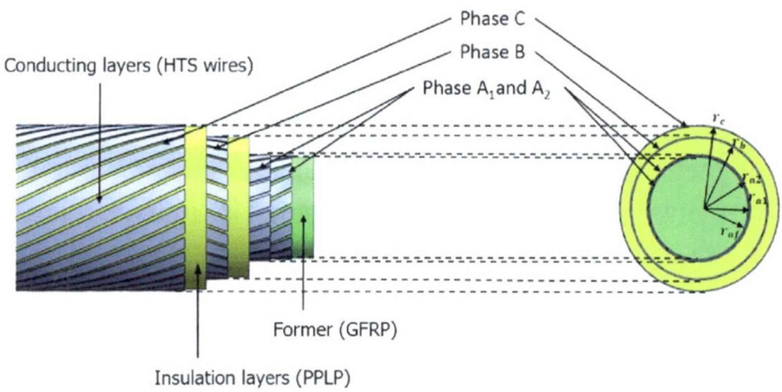

Figure 7.1. *Illustration of structure of the 6-m-long, tri-axial high-temperature superconducting (HTS) power cable.*

Source: Seok-Ju Lee, Creative Commons License.

Tip

Keeping up with innovations in wire fabrication and scalability is crucial for advancing the practical applications of high-temperature superconducting wires.

The utility of HTS materials has been demonstrated in the domain of magnetic levitation (maglev) systems for transportation. The aforementioned materials exhibit remarkable magnetic characteristics, which enable them to exhibit robust attraction or repulsion when subjected to magnetic fields. The utilization of high-temperature superconducting (HTS) magnets in maglev trains facilitates their development, resulting in reduced friction and minimal energy consumption, thereby enabling high-speed transportation. High-temperature superconducting (HTS) maglev systems have the potential to revolutionize urban mobility by providing an environmentally sustainable and highly efficient mode of transportation. This is achieved by levitating trains over the tracks, eliminating the need for traditional mechanical components and reducing energy consumption (Maeda & Yanagisawa, 2013) (Figure 7.1).

Moreover, materials exhibiting high-temperature superconductivity (HTS) are of immense significance in advancing high-field magnet technology. The magnets that produce high-intensity magnetic fields are extensively utilized in scientific investigations and medical diagnoses. High-temperature superconducting (HTS) magnets are indispensable constituents of particle accelerators, nuclear magnetic resonance (NMR) spectrometers, and magnetic resonance imaging (MRI) apparatus. The superior strength and denser composition of high-temperature superconducting (HTS) magnets compared to their conventional counterparts facilitate the attainment of elevated magnetic field intensities and enhanced imaging functionalities. This, in turn, has led to notable progressions in fundamental scientific inquiry and medical diagnostic procedures.

High-temperature superconducting (HTS) materials are of paramount importance in the domain of storing energy and power systems. Superconducting energy storage systems (SMES) that utilize high-temperature superconducting (HTS) materials have been found to possess highly efficient and expeditious energy discharge capabilities. The aforementioned systems can accumulate electrical energy within a magnetic field and subsequently discharge it expeditiously as required. Small and medium-sized enterprise (SME) systems exhibit promising prospects for employment in stabilising the grid, enhancing power quality, and providing backup power supply. HTS-based superconducting fault current limiters (SFCL) play a crucial role in preserving the stability of power grids by rapidly mitigating overcurrents during faults, thereby safeguarding the dependability and stability of power systems (Iijima & Matsumoto, 2000) (Figure 7.2).

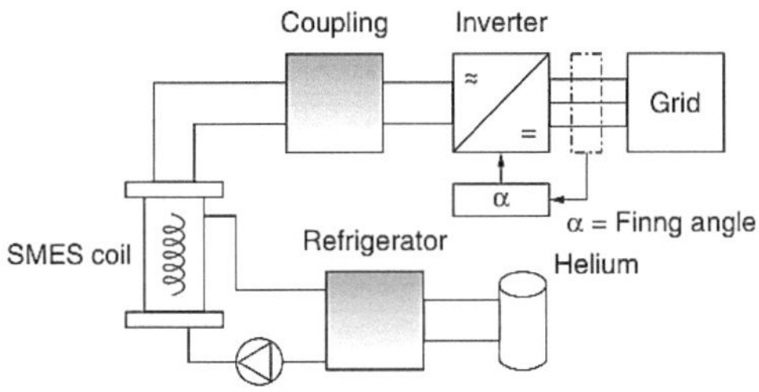

Basic setup of a SMES unit.

Figure 7.2. *Schematic of superconducting magnetic energy storage system.*

Source: The Teche, Creative Commons License.

The ramifications of HTS are also evident in the domains of electronics and computing. Superconducting electronics offer high-speed processing and low energy consumption by functioning at very low temperatures. The aforementioned technology exhibits potential utility in the domains of computing with high performance, signal processing, and communication systems. High-temperature superconducting (HTS) materials exhibit potential in quantum computing, wherein their distinctive characteristics are utilized for generating and manipulating qubits, the fundamental building blocks of quantum information processing. Quantum computing systems based on HTS have the potential to bring about a revolutionary change in computation by facilitating substantially faster and more intricate calculations as to classical computing.

Finally, hydrothermal synthesis (HTS) materials are utilized in developing sensors and detectors that require a high degree of sensitivity. Superconducting sensors, including SQUIDs (superconducting quantum interference devices), demonstrate exceptional precision in determining weak magnetic fields. These phenomena find utility in various scientific domains, including geophysics, astronomy, materials characterization, and other related fields. High-Threshold Scintillation (HTS) detectors are utilized in experiments in high-energy physics and astrophysics research. Detecting uncommon occurrences and particles is of utmost importance in these applications (Noe & Steurer, 2007).

The significance of high-throughput screening (HS) in technological progress is extensive and ongoing. The absence of resistance in conducting electrical currents, coupled with their distinctive magnetic characteristics, has facilitated noteworthy advancements in energy conveyance, transportation, medical diagnosis, computing, and sensing. With advancements in research and manufacturing methods, high-temperature superconducting (HTS) materials have the capacity to significantly transform various industries, enhancing the efficiency and sustainability of our technological infrastructure, while also enabling scientific exploration beyond current limits (Noe & Steurer, 2007).

7.3. SUPERCONDUCTING WIRES AND CABLES

7.3.1. Advantages of High-Temperature Superconducting Wires

The utilization of high-temperature superconducting wires presents a multitude of benefits in contrast to traditional wires, rendering them exceedingly appealing for diverse applications. A notable benefit lies in the elevated critical temperature (Tc) exhibited by these superconductors in contrast to conventional ones. Superconductors with high-temperature properties have the ability to function at temperatures that exceed the boiling point of liquid nitrogen (77 K), thereby enabling the utilization of cryogenic cooling techniques that are more cost-effective and readily available (Schiferl et al., 2006).

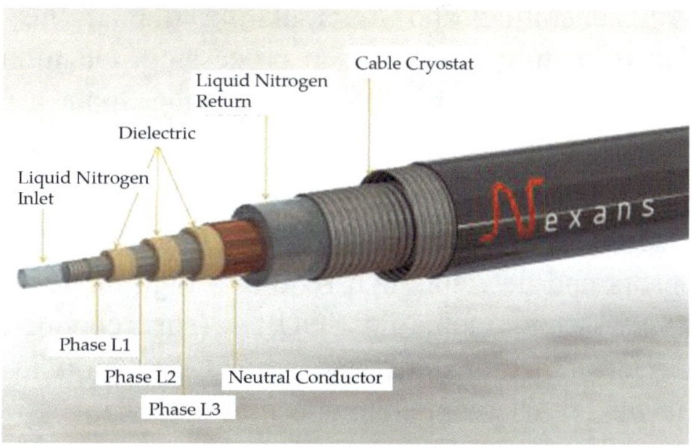

Figure 7.3. *Image of world's longest conducting cable.*

Source: Helmholtz Association, Creative Commons License.

An additional advantage pertains to the notably reduced electrical resistance exhibited by superconducting wires when in superconductivity. Following the critical temperature, these wires demonstrate a complete absence of electrical resistance, thereby facilitating the conduction of electrical current without any power dissipation due to resistance. This particular attribute mitigates energy inefficiencies and expedites the conveyance of greater quantities of electric power across extended distances.

Compared to conventional conductors like copper or aluminum wires, high-temperature superconducting wires exhibit a greater capacity for carrying current.

This implies that they can sustain considerably elevated current densities without undergoing considerable losses or overheating. The heightened current density is ascribed to the distinctive characteristics of high-temperature superconductors, including their capacity to sustain a high critical current density and the existence of flux pinning mechanisms that impede the movement of magnetic flux lines. The aforementioned characteristics facilitate the effective conduction of electrical current within the wire, thereby enabling the conveyance of significant quantities of energy (Xie et al., 2007) (Figure 7.3).

7.3.2. Current Carrying Capacity and Reduced Power Losses

The augmented ability to carry current exhibited by high-temperature superconducting wires is a pivotal characteristics that distinguishes them from traditional conductors. The maximum current density that can be maintained without losing the superconducting state is defined by superconducting wires' critical current density (Jc). Superconductors with high-temperature properties exhibit critical current densities that are relatively elevated, thereby enabling them to transport substantial quantities of electrical current prior to undergoing a resistive phase (Masuda et al., 2005).

The reduced resistive losses exhibited by superconducting wires offer significant benefits in power distribution and transmission systems. Traditional wires, composed of materials like copper or aluminium, undergo resistive losses that lead to the loss of electrical energy as heat. On the contrary, it can be observed that superconducting wires demonstrate an insignificant amount of resistance, leading to minimal power dissipation while being transmitted. This attribute facilitates enhanced power allocation, minimized energy dissipation, and reduced expenses linked with energy dissipation.

Enhanced power transmission efficiency is a direct consequence of minimized power losses. Superconducting wires possess the ability to transport substantial electric currents across extensive distances with minimal energy loss, thereby resulting in reduced voltage drops during the transmission process. The aforementioned characteristics holds significant importance in the context of long-range power transmission, as it minimizes the requirement for expensive voltage regulation apparatus and enhances the overall efficiency of the system (Stovall et al., 2001).

7.3.3. Applications in Power Transmission and Distribution

The utilization of high-temperature superconducting wires offers many advantages, particularly in the realm of power transfer and distribution networks. The augmented capacity of current conduction enables greater power transfer and enhanced transmission capability. The utilization of superconducting wires facilitates the conduction of substantially elevated currents, which allows the transmission of greater quantities of electrical power across the grid. This particular ability is particularly advantageous in situations where power transmission capacity is restricted by existing infrastructure or in areas with high demand.Superconducting wires with high-temperature capabilities have been utilized in diverse power transfer and distribution domains, encompassing extensive distance power transmission, subterranean cables, and offshore turbines. Superconducting cables are a highly efficient means of transmitting electricity over extended distances with little energy dissipation. This characteristics renders them particularly well-suited for linking power grids or transferring electricity from distant generation sources to densely populated regions. In addition, the diminutive dimensions and low mass of superconducting cables may allow for simplified deployment and diminish construction expenses in contrast to traditional power transmission frameworks. Incorporating high-temperature superconducting wires within power distribution networks presents prospects for enhanced grid stability and dependability. The increased capacity of current carrying enables the transmission of greater power loads, thereby mitigating the likelihood of overloading and voltage fluctuations. This capability proves to be particularly advantageous in situations with significant fluctuations in power demand or emergency scenarios that necessitate temporary power rerouting (Yuan et al., 2018).

7.3.4. Innovations in Wire Fabrication and Scalability

Considerable progress has been achieved at high-temperature superconducting wires, enhancing efficiency and expandability. The oxide-powder-in-tube (OPIT) method is a commonly employed fabrication technique for the production of superconducting wires. This method entails utilizing a metallic tube filled with a precursor powder that includes the requisite elements. Subsequently, the tube undergoes a series of controlled heating and cooling procedures to facilitate the conversion of the precursor powder into a superconducting material (Figure 7.4).

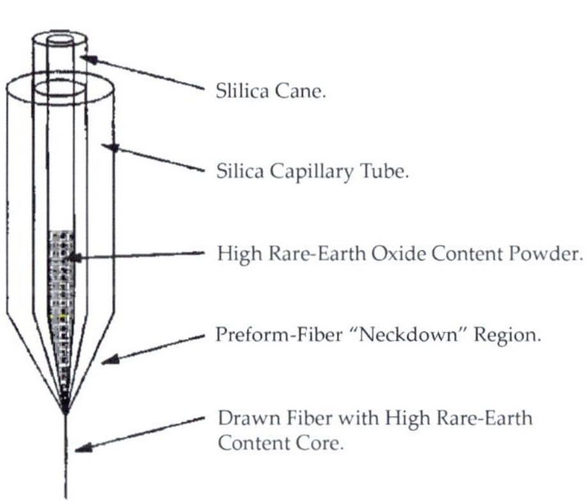

Figure 7.4. *Illustration of powder in tube method.*

Source: John Ballato, Creative Commons License.

Pulsed laser deposition (PLD) is a frequently employed technique in which a laser is utilized to ablate a target material, which is subsequently placed onto a substrate to generate thin films. The employment of pulsed laser deposition (PLD) technique provides a high degree of accuracy in regulating the film composition, thereby facilitating the deposition of superconductive layers on diverse substrates. This, in turn, enables the creation of wires that are both flexible and adaptable (Molodyk et al., 2021) (Figure 7.5).

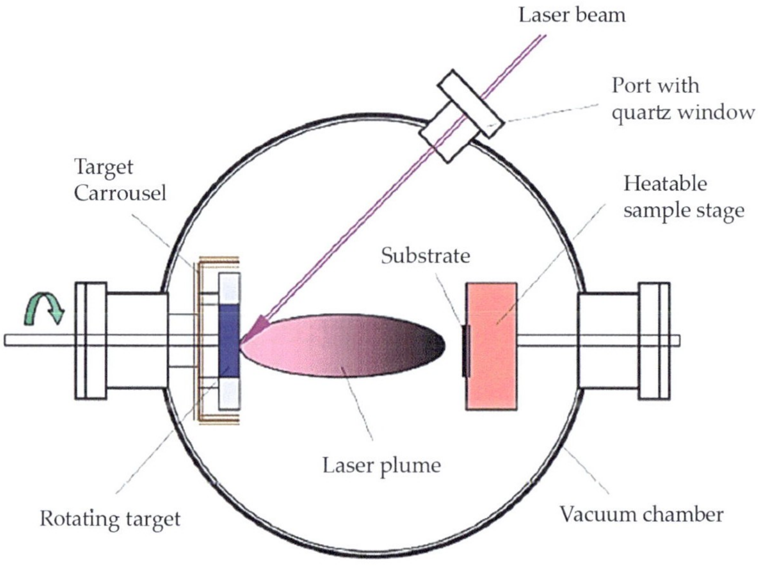

Figure 7.5. *Schematic of pulsed laser deposition (PLD) technique.*

Source: Voccoat, Creative Commons License.

Remember

Superconducting bearings enable frictionless motion in magnetic levitation and transportation systems, offering significant advantages such as reduced energy consumption and maintenance costs.

The development of wire fabrication methods has led to the production of wires that exhibit superior superconducting properties, enhanced mechanical characteristics, and increased critical current densities. Scholars have been engaged in the endeavor of enhancing the configuration and microarchitecture of superconductive substances to attain elevated critical current densities, thereby enabling the conveyance of greater electrical currents.

The actual implementation of high-temperature superconducting wires continues to face the challenge of scalability. The task of achieving consistency and high performance while producing wires of considerable length is a multifaceted undertaking. Current endeavors are being made to establish methodologies that facilitate the production of wires spanning a distance of one kilometer, while maintaining uniform characteristics. Furthermore, the current commercialization endeavors are centered on cost reduction and enhancement of production scalability of high-temperature superconducting wires to increase their economic feasibility for extensive application (Rupich, 2015).

7.4. MAGNETIC LEVITATION AND TRANSPORTATION SYSTEMS

7.4.1. Magnetic Levitation Principles and Advantages

Maglev, short for magnetic levitation, is an innovative mode of transportation based on electromagnetism's fundamental principles. Maglev systems have several advantages over conventional transportation methods because they use magnetic fields to lift and propel vehicles. The absence of mechanical interaction and friction is considered a significant advantage. Maglev trains are designed to levitate above the guideway by using repulsive forces produced between the magnets on the train and the guideway. This unique feature enables the train to be propelled forward while maintaining a floating position. The absence of friction in the motion leads to decreased energy dissipation and system degradation, thereby yielding improved energy efficacy and reduced maintenance needs (Du et al., 2019). Besides its ability to provide motion without friction, maglev technology presents various other benefits. The lack of tactile interaction between the train and the guideway results in a notable reduction in both noise and vibrations, thereby affording passengers a more seamless and agreeable travel experience. Maglev trains can achieve higher speeds than traditional trains because they can levitate over the ground. Maglev trains can attain speeds exceeding 300 miles per hour due to their lower rolling resistance, thereby facilitating expedited and more effective transportation (Li et al., 2022).

7.4.2. High-Temperature Superconductors in Maglev Trains and Transportation

The utilization of high-temperature superconductors has become a crucial element in the advancement of maglev technology. In contrast to traditional superconductors, high-temperature superconductors exhibit the ability to sustain superconductivity at temperatures that surpass the boiling point of liquid nitrogen, which is approximately –196°C. The elevated critical temperature (Tc) of maglev systems permits their operation at comparably higher temperatures, thereby obviating the requirement for costly cryogenic cooling systems. The utilization of high-temperature superconductors streamlines infrastructure prerequisites and diminishes operational expenditures.

High-temperature superconductors exhibit remarkable magnetic characteristics, rendering them highly suitable for implementation in magnetic levitation (maglev) systems. Superconductors demonstrate a notable critical current density (Jc), enabling them to transport substantial electrical currents with negligible energy dissipation. The attainment of a high critical current density (Jc) is paramount in facilitating the production of robust magnetic fields, allowing for the reliable levitation and transportation of maglev trains. In addition, it is noteworthy that high-temperature superconductors exhibit exceptional abilities in terms of flux pinning. The magnetic flux can be efficiently confined and restricted, resulting in heightened stability and safeguarding against potential derailment of the train caused by external perturbations (Deng et al., 2017).

7.4.3. Superconducting Bearings and Frictionless Motion

Implementing superconducting bearings is essential for attaining motion without friction in magnetic levitation transportation systems. The aforementioned bearings use the Meissner phenomenon, manifesting when a superconductor repels magnetic fields from the inside upon entering the superconducting phase. The utilization of superconducting materials in the bearings of maglev trains enables them to levitate above the guideway without any physical contact, thereby creating an environment with negligible friction.

The superconducting bearings collaborate with electromagnetic guidance systems to sustain stability and regulate the levitation of the train. The electromagnetic configuration comprises a series of magnets affixed to the train and a collection of coils on the guideway. The superconducting magnets produce magnetic fields that engage with the electromagnetic coils, thereby regulating the levitation altitude and furnishing stability against both horizontal and vertical disturbances (Kuehn et al., 2007).

The lack of physical contact and resultant friction between the train and guideway confers notable benefits. The reduction of wear and tear on the system leads to an extension of the infrastructure's lifespan and a decrease in maintenance demands. The absence of friction results in reduced energy consumption as there is no dissipation of energy due to mechanical impediments. Maglev trains that utilize superconducting bearings demonstrate elevated energy efficiency and decreased operational expenses (Arsénio et al., 2016).

7.4.4. Future Predictions and Developments in Transportation Technologies

Ongoing research and development efforts are focused on advancing maglev transportation technology, which holds significant promise for the future. Current endeavors are being made to enhance the steadiness and effectiveness of maglev systems while tackling obstacles such as horizontal instability and the management of vibrations. Academic institutions and industry experts are engaged in a collaborative effort to enhance the design of superconducting magnets, create creative levitation control systems, and improve the general efficiency of maglev trains (Harris et al., 2015).

Furthermore, promising opportunities exist for the amalgamation of maglev technology with alternative modes of transportation. The prospect of developing integrated networks that merge maglev trains with traditional rail systems is currently being investigated by scholars to establish a cohesive and effective multimodal transportation framework. Incorporating maglev technology into pre-existing transportation infrastructures can fundamentally transform human transportation, facilitating expedited and expedient travel over extended distances (Bimbraw, 2015).

7.5. SUPERCONDUCTING MAGNETS

7.5.1. High-Field Magnets for Scientific Research and Medical Imaging

The utilization of high-field magnets is paramount in the progression of scientific inquiry and the transformation of medical imaging methodologies. Magnets that employ superconducting materials, particularly high-temperature superconductors, can produce robust and consistent magnetic fields. This property allows for the investigation of fundamental physics events and promotes advancements in the fields of materials science and chemistry (Takeda et al., 2022).

High-field magnets are essential instruments utilized in diverse applications within scientific research. The utilization of magnets in particle accelerators is crucial for the manipulation and concentration of charged particles in experiments pertaining to high-energy physics. This technology enables researchers to investigate subatomic particles, gain insight into fundamental forces, and unravel enigmas surrounding the cosmos. Using high-field magnets allows scientists to conduct investigations into the properties and relationships of matter at an unparalleled degree by facilitating exact control over particle trajectories (Wang et al., 2021).

Did You Know?

Superconducting qubits, based on the principles of superconductivity, are key components in the field of quantum computing, offering the potential for exponential computational power and transformative applications.

Nuclear magnetic resonance (NMR) spectroscopy represents a crucial utilization of high-field magnets. Nuclear magnetic resonance (NMR) spectroscopy is a prevalent analytical technique employed in the fields of chemistry, biochemistry, and medicine. This method employs robust magnetic fields to orient nuclear spins, thereby enabling researchers to obtain valuable information regarding the molecular structure, movement, and relationships of diverse compounds. The utilization of the distinctive characteristics of high-field superconducting magnets has rendered NMR spectroscopy an indispensable instrument for scrutinizing chemical structures, deciphering biomolecular interactions, and facilitating drug discovery and development (Moser et al., 2017) (Figure 7.6).

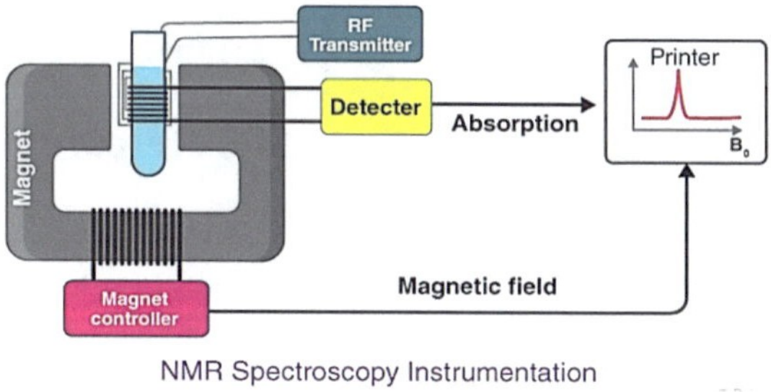

Figure 7.6. *Diagram of nuclear magnetic resonance (NMR) spectroscopy.*

Source: Byju's, Creative Commons License.

High-field magnets, which incorporate superconducting technology, have transformed the application of magnetic resonance imaging (MRI) within the domain of medical imaging. Magnetic Resonance Imaging (MRI) utilizes robust magnetic fields to produce intricate and high-fidelity visual representations of the human anatomy, thereby facilitating the identification and management of diverse pathological states. Superconducting magnets present notable benefits in MRI, delivering more robust and consistent magnetic fields than traditional magnets.

The augmentation of these more potent fields leads to an increase in the ratio of signal to noise, which in turn, yields an amelioration in the quality of the image, a greater degree of spatial resolution, and an enhancement in the contrast of the tissue. Moreover, the constancy of the magnetic field diminishes image distortion and artifacts, resulting in enhanced precision and dependability of diagnoses.

Current investigations in high-field magnet technology concentrate on expanding the limits of magnetic field intensity and investigating novel magnet configurations and substances. Researchers endeavor to attain greater field strengths and more condensed magnets to enhance functionality, expand the scope of potential applications, and optimize cost efficiency. Research is underway to explore advanced magnet designs that utilize high-temperature superconducting materials to achieve stronger, more efficient, and more reliable magnetic fields. Furthermore, the progress made in cryogenic cooling systems and magnet fabrication techniques significantly advances durable and expandable high-field magnet technologies (Gourlay et al., 2004).

7.5.2. Superconducting Magnets in Fusion Energy Research

The utilization of superconducting magnets is of utmost importance in the endeavor to attain fusion energy, which holds promise as an environmentally friendly and inexhaustible energy resource. Fusion energy aims to emulate the fusion process that transpires in the Sun, whereby low-mass atomic nuclei merge to discharge substantial quantities of energy. In order to attain and sustain the requisite circumstances for regulated fusion, potent magnetic fields are imperative for the confinement of the heated plasma (Abdou, 1978).

Superconducting magnets play a crucial role in generating and regulating magnetic fields that are indispensable for plasma confinement in fusion reactors. The magnets generate a magnetic field arrangement that ensnares and confines the high-temperature plasma, facilitating the attainment of the necessary temperature and pressure levels for the initiation of fusion reactions. The magnets' superconductivity guarantees a robust and consistent magnetic field, thereby inhibiting the plasma from making contact with the walls of the fusion chamber. The confinement mechanism facilitates effective energy transfer, thus maintaining the fusion process and facilitating the liberation of substantial energy.

The utilization of high-temperature superconductors has significantly impacted the field of fusion energy research. The aforementioned materials demonstrate elevated critical magnetic fields and enhanced performance at increased temperatures, rendering them highly appropriate for the rigorous conditions of fusion reactors. The progress made in the field of high-temperature superconductors has resulted in the creation of superconducting magnets that are both compact and powerful. This development has contributed to the reduction of the size and intricacy of fusion reactors, while simultaneously enhancing their overall performance and efficiency (Nakajima et al., 1990).

The International Thermonuclear Experimental Reactor (ITER) is a noteworthy fusion endeavor that is dependent on the utilization of superconducting magnets. The primary objective of ITER is to exhibit the scientific and technical viability of fusion energy at a significant magnitude. The project employs superconducting magnets to produce an unparalleled magnetic field, which confines the plasma and facilitates the occurrence of sustained fusion reactions. The realization of practical fusion power, which presents a clean, safe, and practically limitless energy source that can satisfy the world's energy needs, is contingent upon the accomplishment of ITER and other fusion projects that rely on superconducting magnets (Ito et al., 2018).

Current investigations in the field of superconducting magnet technology for fusion energy are centered on advancing high-temperature superconductors and inventive magnet configurations. Researchers are currently investigating innovative materials and configurations with the aim of improving the efficacy and efficiency of fusion magnets. The primary objective is to establish durable, economical, and financially feasible superconducting magnets to facilitate the extensive implementation of fusion energy systems. The ongoing progress in superconducting magnet technology presents potential transformative changes in the global energy sector, with the capacity to tackle pressing issues of energy sustainability and climate change (McHenry & Reed, 1980).

7.6. ENERGY STORAGE AND SUPERCONDUCTING FAULT CURRENT LIMITERS

Integrating energy storage systems is paramount in mitigating the challenges posed by the fluctuating and unpredictable nature of renewable energy sources, and in catering to the escalating requirements of contemporary power networks. Superconducting energy storage systems (SMES) possess distinct benefits with regard to their high energy density, rapid response rates, and minimal energy dissipation. The application of high-temperature superconductors (HTS) in superconducting magnetic energy storage (SMES) has significantly transformed the field of energy storage technology.

The operational mechanism of SMES involves the induction of direct current throughout a superconducting coil, thereby producing a potent magnetic field and accumulating electrical energy in the configuration of magnetic energy. The high critical current density exhibited by high-temperature superconducting (HTS) materials facilitates the effective storage of substantial quantities of energy within a confined area. In contrast to traditional energy storage technologies, Superconducting Magnetic Energy Storage (SMES) demonstrates superior energy densities, accelerated charging and discharging rates, and extended cycle lifetimes. Superconducting Magnetic Energy Storage (SMES) systems based on High Temperature Superconductors (HTS) have been utilized in diverse fields. Implementing SMES in grid stabilization can offer dynamic energy reimbursement and voltage support, thereby contributing to the preservation of grid stability and alleviating power fluctuations. The grid's stability is ensured through the rapid response of SMES systems to frequency deviations, making frequency regulation a crucial application (Moon et al., 2013).

The short-term availability of renewable sources poses a significant challenge in integrating renewable energy. Small and medium-sized enterprises (SMEs) can accumulate surplus energy during heightened generation phases and discharge it during low-generation periods, thereby mitigating fluctuations and enhancing the grid's dependability. Moreover, SMES assumes a crucial function in the context of uninterruptible power supply (UPS) systems, by furnishing auxiliary power during periods of power failure, thereby averting data loss or harm to essential infrastructure. The utilization of high-temperature superconducting (HTS) materials serves to augment the efficacy of superconducting magnetic energy storage (SMES) systems. Superconducting HTS tapes and coils demonstrate negligible power dissipation owing to their nearly zero electrical resistance in the superconducting phase. This enables the achievement of enhanced energy retrieval and storage efficacy, thereby diminishing total energy consumption and elevating system efficiency (Guo et al., 2012).

The utilization of high-temperature superconductors (HTS) in power systems finds significant application in the form of superconducting fault current limiters (SFCL). The occurrence of fault currents, which may arise from short circuits or equipment malfunctions, can harm the various components of a power system. SFCLs that utilize high-temperature superconducting (HTS) materials can restrict the amplitude of fault currents, thereby safeguarding the power system from harm and preserving its stability. The functioning of Superconducting Fault Current Limiters (SFCLs) is predicated upon the distinctive characteristics of High-Temperature Superconducting (HTS) materials. When a fault occurs, the high-temperature superconductor material swiftly transitions from a superconducting state to a resistive state, thereby hindering the flow of too many currents. The feature of the current limitation serves to safeguard power system equipment and avert interruptions to the power supply (Chen et al., 2016).

Superconducting Fault Current Limiters (SFCLs) have a diverse range of applications, including but not limited to distribution and transmission networks, renewable energy integration, and industrial power systems. Implementing these devices results in an improvement in the stability of the power grid, a decrease in the occurrence of system downtime, and an overall increase in the dependability of power systems. Current research in Superconducting Fault Current Limiter (SFCL) technology is centered on enhancing the efficiency and response times of High-Temperature Superconducting (HTS) based fault current limiters by optimizing their design and performance. Besides energy storage and fault current limiters, smart grid technologies are also incorporating high-temperature superconductors. Smart grids utilize sophisticated control, monitoring, and communication functionalities to optimize the flow of electricity, enhance efficiency, and bolster the power grid's resilience. High-temperature superconducting (HTS) cables can facilitate the transmission of electrical power with minimal energy losses over extended distances, thereby enhancing the efficiency of the grid (Zhang et al., 2020).

Transformers that utilize high-temperature superconducting (HTS) materials provide benefits such as decreased dimensions, mass, and energy dissipation in contrast to traditional transformers. Because of this, compact and effective distribution systems are made possible, increasing the overall effectiveness of smart grids. Furthermore, superconductors are essential components in systems for monitoring and controlling electrical grids. The integration of High-Temperature Superconducting (HTS) sensors, including Superconducting Fault Current Limiters (SFCLs) and Magnetic Field Sensors (MFSs), can enable smart grids to attain instantaneous monitoring of power transmission, fault identification, and system safeguarding.

Incorporating high-temperature superconductors in smart grid systems offers a multitude of prospects for heightened grid intelligence, refined energy administration, and grid steadiness. Continuing research endeavors to enhance the optimization of the configuration, production, and implementation of HTS reliant components in intelligent power grids, thereby facilitating the development of power systems that are more effective and robust (Dondapati et al., 2017).

7.7. SUPERCONDUCTING ELECTRONICS AND QUANTUM COMPUTING

7.7.1. Development of Superconducting Electronics

The field of superconducting electronics has undergone substantial advancements since its inception, characterized by notable milestones that have delineated its trajectory. The origins of superconducting electronics can be traced back to 1911, when Heike Kamerlingh Onnes first discovered the phenomenon of superconductivity. The initial accomplishments in this domain encompassed the creation of superconducting quantum interference devices (SQUIDs) during the 1960s, which subsequently emerged as an indispensable instrument for precise magnetic field assessments.

Over the course of the subsequent decades, notable progress was achieved in comprehending the characteristics of superconductive substances and their prospective utility in the realm of electronics. The field underwent a further revolution by identifying high-temperature superconductivity in the latter part of the 1980s. The discovery has expanded the potential for practical application owing to these substances' comparatively elevated critical temperatures in contrast to traditional superconductors (Silver et al., 2003).

Superconducting electronics is dependent on the distinctive characteristics of superconducting substances, which manifest an absolute absence of electrical resistance and ideal diamagnetism when their temperatures fall below the critical point. Various superconducting materials have been employed, comprising low-temperature superconductors such as niobium and niobium-titanium alloys, alongside high-temperature superconductors such as yttrium-barium-copper oxide (YBCO) and bismuth strontium calcium copper oxide (BSCCO).

The actualization of superconducting circuits has been made possible through the pivotal role played by the advancement of fabrication techniques. Initially, the primary emphasis was on the implementation of thin-film deposition techniques, including sputtering and molecular beam epitaxy, to fabricate superconducting layers that exhibit superior quality and regulated characteristics. The utilization of these methodologies facilitated the amalgamation of superconductive substances with diverse constituents and substances within electronic circuits in device miniaturization and integration (Tolpygo et al., 2007). Advancements in fabrication techniques and materials engineering have facilitated the reduction of the dimensions of superconducting devices, rendering them more condensed and adaptable to diverse applications. The process of miniaturization has enabled the amalgamation of numerous components onto a solitary chip, resulting in heightened functionality and superior performance.

In recent years, there has been a notable emphasis on amalgamating superconducting devices with traditional electronics. Hybrid systems have been created for applications in high-speed computing, processing of signals, and communication by integrating the distinctive characteristics of superconducting elements with the rapidity and functionality of conventional electronic components (Rowell, 1999).

7.7.2. Applications in High-Speed Computing and Data Transmission

The utilization of superconducting electronics presents notable advantages in high-speed computing, facilitating advancements in processing capabilities and data transmission. One of the primary benefits pertains to the capacity to function at exceedingly low temperatures, wherein superconductive substances demonstrate a complete absence of electrical resistance. This attribute eliminates the power dissipation that is usually linked with resistance in traditional electronics, thereby resulting in a decrease in heat production and an enhancement in energy efficiency. Consequently, the utilization of superconducting electronics can attain elevated clock frequencies and computational speeds, rendering them optimal for high-speed computing implementations.

Superconducting electronics offer significant advantages in the domain of digital circuits and logic gates. Josephson junctions are utilized in superconducting digital circuits, exhibiting a distinctive shift from a superconducting state to a resistive state upon reaching a particular current or voltage threshold. The aforementioned characteristics facilitates the development of high-speed switching mechanisms, resulting in enhanced processing rates and decreased latency. Superconducting logic gates, exemplified by the Rapid Single Flux Quantum (RSFQ) circuits, can function at clock frequencies that surpass 100 GHz, thereby outperforming conventional CMOS-based technologies (Sivakumar et al., 2000).

The potential of utilizing superconducting analog and mixed-signal circuits in high-speed computing applications is noteworthy. The circuits in question employ superconducting elements, specifically superconducting quantum interference devices (SQUIDs), to execute the processing of analog signals with unparalleled precision and sensitivity. Due to their low noise properties and superior dynamic range, they are well-suited for various applications such as high-resolution signal processing, ultrafast gathering of data, and low-noise amplification. In addition, the amalgamation of superconducting digital and analog circuits within mixed-signal systems facilitates the smooth integration and effective processing of digital and analog signals.

In the domain of high-performance computing and data centers, superconducting electronics offer a path to overcome the limitations of traditional computing architectures. The exceptional processing power and energy efficiency of superconducting technologies make them well-suited for complex calculations, simulations, and data-intensive tasks. Moreover, the inherent scalability of superconducting circuits enables the construction of large-scale systems with a high density of computational elements, facilitating the realization of exascale computing capabilities. This potential breakthrough has significant implications for fields such as scientific research, weather modeling, cryptography, and artificial intelligence.

Utilizing superconducting interconnects presents a viable remedy for transmitting high-bandwidth data, a critical requirement for contemporary computing and communication systems. The conventional copper interconnects are subject to signal loss and bandwidth restrictions due to resistance and capacitive phenomena. On the contrary, superconducting interconnects demonstrate an almost negligible amount of resistance, thereby facilitating the transfer of signals without any substantial loss or distortion. This characteristics enables the achievement of rapid data transmission with exceptionally elevated data transfer speeds, thereby expediting the effective conveyance of substantial quantities of data within computational frameworks, data hubs, and even over extended distances (Suresh et al., 2010).

7.7.3. Superconducting Qubits and Quantum Computing

Quantum computing is a groundbreaking computational paradigm that utilizes the principles of quantum mechanics to execute computations that surpass the capabilities of classical computers. The qubit, the basic component of quantum information, is at the core of quantum computing. In contrast to classical bits that can represent only a 0 or a 1, qubits possess the unique characteristics of being able to be present in a superposition of both states concurrently. This property allows for the utilization of quantum parallelism. Due to their scalability, controllability, and extended coherence times, superconducting qubits have become a prominent choice for qubit implementation.

1. **Kinds of Superconducting Qubits (e.g., Flux, Charge, Phase Qubits):** Superconducting qubits are available in diverse forms, each exploiting distinct physical characteristics to encode and handle quantum information. Using the flux quantum as the foundation for its operations characterizes flux qubits, which also depend on Josephson junctions to generate and regulate the qubit states. In contrast to other types of qubits, charge qubits leverage the distinct charge of Cooper pairs to represent quantum

information. This is achieved by utilizing the unique characteristics of superconducting islands and tunnel junctions. The encoding of qubit states in phase qubits is achieved by utilizing the quantum phase disparity that exists across Josephson junctions. The diverse array of qubit variants affords researchers the opportunity to investigate various quantum computing methodologies, as they offer flexibility in their structure and performance attributes (Xue & Wang, 2007).

2. **Quantum Coherence and Decoherence in Superconducting Systems:** Maintaining quantum coherence, which facilitates the superposition and involvement of qubit states, is a significant obstacle in implementing superconducting qubits. The coherence of quantum systems can be influenced by diverse sources of noise, including fluctuations in the ambient environment, processes of energy relaxation, and dephasing. The phenomenon of decoherence occurs due to the interaction between the intricate quantum states and their surrounding environment, resulting in the gradual dissipation of coherence. Scientists utilize diverse methodologies, including cryogenic cooling and correction of quantum error codes, to alleviate the impacts of decoherence and prolong the coherence durations of superconducting qubits.
3. **Quantum Gates and Quantum Operations Using Superconducting Qubits:** In order to execute computational operations on quantum computers, the utilization of quantum gates is necessary to manipulate the states of qubits and produce entanglement. Quantum gates can be implemented on superconducting qubits using tailored electromagnetic fields and microwave pulses. Frequently employed gates in quantum computing comprise the Hadamard gate, Pauli gates (X, Y, Z), and two-qubit gates such as the controlled-NOT (CNOT) gate. Superconducting qubits can execute intricate quantum algorithms by precisely regulating the timing and parameters associated with gate operations (Chow et al., 2012).
4. **Superconducting Qubit Architectures (e.g., Transmon, Xmon, Gmon):** Various architectures have been devised to enhance the efficiency and expandability of superconducting qubits. The transmon qubit has been engineered to possess a resilient energy spectrum, thereby mitigating the impact of charge noise and amplifying coherence durations. The Xmon qubit is designed

with supplementary Josephson junctions to enhance coherence and enable interaction with adjacent qubits. The gmon is a type of transmon qubit that exhibits amplified coupling to microwave resonators, with the "g" in its name denoting "giga" in reference to the gigahertz frequencies involved. The architectures in question undergo a continuous process of refinement and customization in order to attain extended coherence times, elevated gate fidelities, and improved scalability.

The enhancement of superconducting qubit performance is a significant area of interest in the realm of research and development, with various challenges and approaches being explored. The primary obstacles to overcome involve mitigating noise and decoherence, enhancing qubit lifespans and gate fidelities, and resolving cross-talk and scalability concerns. Addressing these obstacles includes creating innovative materials and manufacturing methods, utilizing advanced control and measurement methodologies, and integrating quantum correction codes. The objective of collaborative endeavors among academia, industry, and government entities is to propel the field forward and establish a path towards developing fault-tolerant quantum computers that can resolve intricate problems and transform diverse domains of science and technology (Calzona & Carrega, 2022).

7.7.4. Advancements and Challenges in the Field of Superconducting Electronics

The domain of superconducting electronics has witnessed noteworthy advancements in high-performance computing and quantum technologies, albeit with certain challenges. This section explores the complexities of recent advancements and the diverse obstacles faced in this swiftly developing domain.

Significant advancements in the field of superconducting electronics have been observed in recent research efforts. Significant progress pertains to the emergence of sophisticated circuit design methodologies that are custom-made for superconducting circuits. The distinctive characteristics of superconducting materials, including their zero resistance and rapid switching times, is employed in developing circuits that function at elevated speeds and display reduced power dissipation. Significant progress has been achieved by researchers in the optimization of circuit layouts, the implementation of inventive logic gates, and the investigation of new methodologies, such as the utilization of Josephson junctions and flux qubits. The aforementioned endeavors have facilitated the development of superconducting circuits that are more intricate and adaptable, exhibiting superior functionality and performance (Chang et al., 2022).

The operation of superconducting electronics is significantly facilitated by the pivotal role played by cryogenic cooling and wrapping considerations. The necessity of cryogenic cooling systems has become crucial due to the fact that superconducting devices generally operate at exceedingly low temperatures, often in close proximity to absolute zero. Advanced refrigeration techniques, including dilution refrigeration and adiabatic demagnetization refrigeration, have been devised to achieve and sustain desired temperature conditions effectively. Furthermore, the packaging of superconducting circuits necessitates scrupulous attention to detail to safeguard them against external electromagnetic interference and guarantee appropriate thermal management to ensure dependable functionality. Efforts are being made to effectively tackle these challenges by exploring innovative packaging solutions that incorporate low-loss interconnects, robust shielding, and efficient thermal insulation (Devoret & Schoelkopf, 2013).

The implementation of superconducting electronics in large-scale systems presents significant obstacles in terms of scalability and integration. As the circuits become more intricate and larger, the task of preserving coherence and reducing cross-talk becomes progressively challenging. To surmount these obstacles, it is imperative to possess a comprehensive comprehension and mastery of pivotal elements such as the caliber of the materials, the density of defects, and the consistency maintained during the entirety of the manufacturing procedure. Scientists and professionals in the field are consistently improving manufacturing methodologies, enhancing the growth processes of materials, and innovating lithography and patterning techniques in order to facilitate the production of expansive superconductive circuits. The aforementioned endeavors are of utmost importance in achieving tangible implementations, such as quantum computing systems featuring a significant quantity of qubits and complex superconducting logic circuits (Anders et al., 2010).

The readout and control electronics that operate at cryogenic temperatures are a crucial element of superconducting electronic systems. The aforementioned electronic devices facilitate accurate quantification and handling of superconducting qubits and other related apparatus. The fundamental research focus is centered on accomplishing high-fidelity readout and effective control operations, while reducing thermal noise and signal loss. Significant progress has been achieved in advancing cryogenic readout methodologies, including quantum-limited amplifiers and low-noise detectors, that facilitate precise and sensitive measures of the quantum states of superconducting circuits. Furthermore, scholars are investigating novel control methodologies and deploying sophisticated control electronics to guarantee accurate and dependable influence of superconducting qubits and other quantum devices.

The advancement of superconducting electronics is contingent upon the collaboration among academia, industry, and government. Superconducting electronics' intricate and multidisciplinary characteristics demand cooperative endeavors to address the scientific and technological obstacles implicated. Academic research institutions play a significant role in generating fundamental knowledge by means of theoretical modeling, material evaluation, and testing. Industry partners contribute their specialized knowledge in fabrication techniques, system integration, and commercialization elements, the translation of research findings into practical applications. In addition, the provision of financial resources, the establishment of infrastructure, and the implementation of policy measures by the government are crucial in creating a favorable atmosphere for the progress of superconducting electronics technology. Collaborations at the national and international levels, partnerships between academia and industry, and platforms for sharing knowledge all contribute to creating synergies that expedite progress and foster innovation within the field (Yazdani-Asrami et al., 2022).

Activity 7.1

You are an engineer working for a renewable energy company. Develop a design challenge for your colleagues, tasking them with integrating high-temperature superconductors into an existing power transmission system to enhance its efficiency and reduce energy losses.

7.8. SUMMARY

Chapter seven explores the practical applications of high-temperature superconductors across various fields. It begins with an introduction to the chapter's topics and highlights the significance of high-temperature superconductors in technological advancements. The chapter focuses on several practical applications, including superconducting wires and cables, magnetic levitation and transportation systems, superconducting magnets, energy storage, superconducting electronics, and quantum computing. It discusses the advantages of high-temperature superconducting wires in terms of current carrying capacity, reduced power losses, and applications in power transmission and distribution. It explores the principles, advantages, and future predictions of magnetic levitation and transportation systems utilizing high-temperature superconductors. The chapter also covers the use of high-temperature superconducting magnets in scientific research, medical imaging, and fusion energy research. It examines the applications of high-temperature superconductors in energy storage and superconducting fault current limiters. Furthermore, it delves into the development of superconducting electronics, including their applications in high-speed computing, data transmission, and the emerging field of quantum computing.

REVIEW QUESTIONS

1. Discuss the significance of high-temperature superconductors in technological advancements.
2. Explain the advantages of high-temperature superconducting wires and their applications.
3. How do high-temperature superconductors contribute to power transmission and distribution systems?
4. Describe the principles and advantages of magnetic levitation systems utilizing high-temperature superconductors.
5. How do superconducting bearings enable frictionless motion in transportation systems?
6. Discuss the applications of high-temperature superconducting magnets in scientific research and medical imaging.
7. Explain the role of superconducting magnets in fusion energy research.
8. How do high-temperature superconductors contribute to energy storage and superconducting fault current limiters?
9. Describe the development and applications of superconducting electronics.
10. How are high-temperature superconductors utilized in high-speed computing and data transmission?

REFERENCES

1. Abdou, M. A. (1978). Radiation considerations for superconducting fusion magnets. *Journal of Nuclear Materials*, *72*(1–2), 147–167.
2. Anders, S., Blamire, M. G., Buchholz, F. I., Crété, D. G., Cristiano, R., Febvre, P., & Zorin, A. B. (2010). European roadmap on superconductive electronics – status and perspectives. *Physica C: Superconductivity, 470*(23–24), 2079–2126.
3. Arsénio, A. J., Carvalho, M. V., Cardeira, C., Branco, P. C., & Melício, R. (2016, September). Viability of a frictionless bearing with permanent magnets and HTS bulks. In *2016 IEEE International Power Electronics and Motion Control Conference (PEMC)*. IEEE. (Vol. 3, pp. 1231–1236).
4. Bimbraw, K. (2015, July). Autonomous cars: Past, present and future a review of the developments in the last century, the present scenario and the expected future of autonomous vehicle technology. In *2015 12th International Conference on Informatics in Control, Automation and Robotics (ICINCO)*. IEEE. (Vol. 1, pp. 191–198).
5. Calzona, A., & Carrega, M. (2022). Multi-mode architectures for noise-resilient superconducting qubits. *Superconductor Science and Technology*, *36*(2), 023001.
6. Chang, C. C. W., Ding, T. J., Ping, T. J., Chao, K. C., & Bhuiyan, M. A. S. (2022). Getting more from the wind: Recent advancements and challenges in generators development for wind turbines. *Sustainable Energy Technologies and Assessments*, *53*, 102731.
7. Chen, L., Tu, X., Chen, H., Yang, J., Wu, Y., Shu, X., & Ren, L. (2016). Technical evaluation of superconducting fault current limiters used in a micro-grid by considering the fault characteristics of distributed generation, energy storage and power loads. *Energies*, *9*(10), 769.
8. Chow, J. M., Gambetta, J. M., Corcoles, A. D., Merkel, S. T., Smolin, J. A., Rigetti, C., & Steffen, M. (2012). Universal quantum gate set approaching fault-tolerant thresholds with superconducting qubits. *Physical Review Letters, 109*(6), 060501.
9. Deng, Z., Zhang, W., Zheng, J., Wang, B., Ren, Y., Zheng, X., & Zhang, J. (2017). A high-temperature superconducting maglev-evacuated tube transport (HTS Maglev-ETT) test system. *IEEE Transactions on Applied Superconductivity*, *27*(6), 1–8.
10. Devoret, M. H., & Schoelkopf, R. J. (2013). Superconducting circuits for quantum information: an outlook. *Science*, *339*(6124), 1169–1174.
11. Dondapati, R. S., Kumar, A., Kumar, G. R., Usurumarti, P. R., &

Dondapati, S. (2017). Superconducting magnetic energy storage (SMES) devices integrated with resistive type superconducting fault current limiter (SFCL) for fast recovery time. *Journal of Energy Storage*, *13*, 287–295.

12. Du, Y., Zheng, J., Bao, Y., Chen, Y., Yang, X., Hu, Z., & Deng, Z. (2019). Feasibility study of a DC linear motor based on the magnet track of high-temperature superconducting maglev. *IEEE Transactions on Applied Superconductivity*, *30*(3), 1–5.
13. Gourlay, S. A., Sabbi, G., Kircher, F., Martovetsky, N., & Ketchen, D. (2004). Superconducting magnets and their applications. *Proceedings of the IEEE*, *92*(10), 1675–1687.
14. Guo, W., Xiao, L., & Dai, S. (2012). Enhancing low-voltage ride-through capability and smoothing output power of DFIG with a superconducting fault-current limiter – magnetic energy storage system. *IEEE Transactions on Energy Conversion*, *27*(2), 277–295.
15. Harris, I., Wang, Y., & Wang, H. (2015). ICT in multimodal transport and technological trends: Unleashing potential for the future. *International Journal of Production Economics*, *159*, 88–103.
16. Hull, J. R. (2003). Applications of high-temperature superconductors in power technology. *Reports on Progress in Physics*, *66*(11), 1865.
17. Iijima, Y., & Matsumoto, K. (2000). High-temperature-superconductor coated conductors: technical progress in Japan. *Superconductor Science and Technology*, *13*(1), 68.
18. Ito, S., Hashizume, H., Yanagi, N., & Tamura, H. (2018). Advanced high-temperature superconducting magnet for fusion reactors: Segment fabrication and joint technique. *Fusion Engineering and Design*, *136*, 239–246.
19. Kuehn, L., Mueller, M., Schubert, R., Beyer, C., De Haas, O., & Schultz, L. (2007). Static and dynamic behavior of a superconducting magnetic bearing using YBCO bulk material. *IEEE Transactions on Applied Superconductivity*, *17*(2), 2079–2082.
20. Li, H., Ke, Z., Huang, H., Li, Y., Wang, S., & Deng, Z. (2022). Vibration suppression of high-temperature superconducting maglev system via semi-active suspension system. *Physica C: Superconductivity and Its Applications*, *601*, 1354109.
21. Maeda, H., & Yanagisawa, Y. (2013). Recent developments in high-temperature superconducting magnet technology. *IEEE Transactions on Applied Superconductivity*, *24*(3), 1–12.
22. Masuda, T., Yumura, H., Watanabe, M., Takigawa, H., Ashibe, Y., Suzawa, C., & Isojima, S. (2005). High-temperature superconducting cable

technology and development trends. *SEI Technical Review*, *59*(7), 13–19.

23. McHenry, H. I., & Reed, R. P. (1980). Structural alloys for superconducting magnets in fusion energy systems. *Nuclear Engineering and Design*, *58*(2), 219–236.
24. Molodyk, A., Samoilenkov, S., Markelov, A., Degtyarenko, P., Lee, S., Petrykin, V., & Vasiliev, A. (2021). Development and large volume production of extremely high current density YBa2Cu3O7 superconducting wires for fusion. *Scientific reports*, *11*(1), 2084.
25. Moon, W. S., Won, J. N., Huh, J. S., & Kim, J. C. (2013). A study on the application of a superconducting fault current limiter for energy storage protection in a power distribution system. *IEEE Transactions on Applied Superconductivity*, *23*(3), 5603404–5603404.
26. Moser, E., Laistler, E., Schmitt, F., & Kontaxis, G. (2017). Ultra-high field NMR and MRI—the role of magnet technology to increase sensitivity and specificity. *Frontiers in Physics*, *5*, 33.
27. Muzzi, L., De Marzi, G., Di Zenobio, A., & Della Corte, A. (2015). Cable-in-conduit conductors: lessons from the recent past for future developments with low and high temperature superconductors. *Superconductor Science and Technology*, *28*(5), 053001.
28. Nakajima, H., Yoshida, K., & Shimamoto, S. (1990). Development of new cryogenic steels for the superconducting magnets of the fusion experimental reactor. *ISIJ International*, *30*(8), 567–578.
29. Noe, M., & Steurer, M. (2007). High-temperature superconductor fault current limiters: concepts, applications, and development status. *Superconductor Science and Technology*, *20*(3), R15.
30. Rowell, J. M. (1999). Recommended directions of research and development in superconducting electronics. *IEEE Transactions on Applied Superconductivity*, *9*(2), 2837–2848.
31. Rupich, M. W. (2015). Second-generation (2G) coated high-temperature superconducting cables and wires for power grid applications. In *Superconductors in the Power Grid*. Woodhead Publishing. (Vol. 2, pp. 97–130).
32. Schiferl, R., Flory, A., Livoti, W. C., & Umans, S. D. (2006, September). High temperature superconducting synchronous motors: Economic issues for industrial applications. In *2006 Record of Conference Papers-IEEE Industry Applications Society 53rd Annual Petroleum and Chemical Industry Conference*. IEEE. (Vol. 1, pp. 1–9).
33. Shen, B., Grilli, F., & Coombs, T. (2020). Overview of H-formulation: A versatile tool for modeling electromagnetics in high-temperature

superconductor applications. *IEEE Access*, *8*, 100403–100414.

34. Silver, A., Kleinsasser, A., Kerber, G., Herr, Q., Dorojevets, M., Bunyk, P., & Abelson, L. (2003). Development of superconductor electronics technology for high-end computing. *Superconductor Science and Technology*, *16*(12), 1368.
35. Sivakumar, H., Bailey, S., & Grossman, R. L. (2000, November). PSockets: The case for application-level network striping for data intensive applications using high speed wide area networks. In *SC'00: Proceedings of the 2000 ACM/IEEE Conference on Supercomputing*. IEEE. (Vol. 4, pp. 38–38).
36. Stovall, J. P., Demko, J. A., Fisher, P. W., Gouge, M. J., Lue, J. W., Sinha, U. K., & Tolbert, J. C. (2001). Installation and operation of the Southwire 30-meter high-temperature superconducting power cable. *IEEE Transactions on Applied Superconductivity*, *11*(1), 2467–2472.
37. Suresh, J., Srinivasan, A., & Damodaram, A. (2010, March). Performance analysis of various high speed data transfer protocols for streaming data in long fat networks. In *2010 International Conference on Recent Trends in Information, Telecommunication and Computing*. IEEE. (Vol. 5, pp. 234–237).
38. Takeda, Y., Maeda, H., Ohki, K., & Yanagisawa, Y. (2022). Review of the temporal stability of the magnetic field for ultra-high field superconducting magnets with a particular focus on superconducting joints between HTS conductors. *Superconductor Science and Technology*, *35*(4), 043002.
39. Tolpygo, S. K., Yohannes, D., Hunt, R. T., Vivalda, J. A., Donnelly, D., Amparo, D., & Kirichenko, A. F. (2007). Process Development for Superconducting Integrated Circuits With 80 GHz Clock Frequency. *IEEE Transactions on Applied Superconductivity*, *17*(2), 946–951.
40. Wang, Q., Liu, J., Zheng, J., Qin, J., Ma, Y., Xu, Q., & Liu, F. (2021). Progress of ultra-high-field superconducting magnets in China. *Superconductor Science and Technology*, *35*(2), 023001.
41. Weiss, J. D., Mulder, T., ten Kate, H. J., & van der Laan, D. C. (2016). Introduction of CORC® wires: highly flexible, round high-temperature superconducting wires for magnet and power transmission applications. *Superconductor Science and Technology*, *30*(1), 014002.
42. Xie, Y. Y., Tekletsadik, K., Hazelton, D., & Selvamanickam, V. (2007). Second generation high-temperature superconducting wires for fault current limiter applications. *IEEE Transactions on Applied Superconductivity*, *17*(2), 1981–1985.
43. Xue, Z. Y., & Wang, Z. D. (2007). Simple unconventional geometric

scenario of one-way quantum computation with superconducting qubits inside a cavity. *Physical Review A*, *75*(6), 064303.

44. Yazdani-Asrami, M., Sadeghi, A., Song, W., Madureira, A., Murta-Pina, J., Morandi, A., & Parizh, M. (2022). Artificial intelligence methods for applied superconductivity: Material, design, manufacturing, testing, operation, and condition monitoring. *Superconductor Science and Technology*, *35*(12), 123001.
45. Yuan, W., Venuturumilli, S., Zhang, Z., Mavrocostanti, Y., & Zhang, M. (2018). Economic feasibility study of using high-temperature superconducting cables in UK's electrical distribution networks. *IEEE Transactions on Applied Superconductivity*, *28*(4), 1–5.
46. Zhang, G., Wang, H., Qiu, Q., Zhang, Z., Xiao, L., & Lin, L. (2020). Recent progress of superconducting fault current limiter in China. *Superconductor Science and Technology*, *34*(1), 013001.
47. Zhao, Y., Zhu, J. M., Jiang, G. Y., Chen, C. S., Wu, W., Zhang, Z. W., & Yamada, Y. (2019). Progress in fabrication of second generation high temperature superconducting tape at Shanghai Superconductor Technology. *Superconductor Science and Technology*, *32*(4), 044004.

CHAPTER

8 Emerging Research and Innovative Directions in High-Temperature Superconductivity

Contents

8.1. INTRODUCTION

The phenomenon of superconductivity is a noteworthy occurrence in which specific materials, upon being cooled at less than a critical temperature, exhibit the ability to conduct electrical current with zero resistance. Historically, the phenomenon of superconductivity has been detected exclusively at temperatures that are exceedingly low, in close proximity to absolute zero. During the mid-1980s, a significant advancement was made in the field of superconductivity whereby it was discovered that specific materials could demonstrate superconductivity at temperatures that were previously deemed unattainable. This discovery had a revolutionary impact on the field. The aforementioned substances are commonly referred to as superconductors that undergo high temperatures. The term "high-temperature superconductivity" pertains to the occurrence of superconductivity at temperatures surpassing the boiling point residing in liquid nitrogen (–196°C or 77K). This temperature threshold is notably higher than that necessary for regular superconductors. The unearthing of high-temperature superconducting (HTS) materials, including the cuprates based on copper along with pnictides based upon iron, has presented novel prospects for pragmatic implementations and revitalized scholarly curiosity in comprehending the fundamental mechanisms that underlie superconductivity (Zareapour et al., 2012; Pamidi et al., 2015).

The significance of emerging studies and novel directions within high-temperature superconductive properties is multifaceted and noteworthy. Initially, it can be posited that high-temperature superconducting (HTS) materials possess the capacity to fundamentally transform diverse technological domains, such as energy transfer and storage, electronic devices, and medical imaging. The utilization of high-temperature superconducting (HTS) materials has the potential to facilitate lossless electricity transfer and effective energy storage, thereby offering a promising pathway toward a sustainable as well as energy-efficient future. In addition, despite significant advancements in high-temperature superconductive properties (HTS) studies throughout recent decades, several fundamental inquiries regarding the characteristics of this phenomenon remain unresolved. The comprehension of the processes that enable specific materials to demonstrate superconductive properties at relatively elevated temperatures remains restricted. Investigating these mechanisms along with elucidating the enigmas of high-temperature superconductivity may result in significant advancements in the fields of condensate matter physics as well as quantum materials (Wang et al., 2019).

In addition, novel avenues in HS investigations have the potential to unveil novel high-temperature superconducting materials that exhibit superior properties and elevated critical temperatures. The investigation of new materials as well as their properties is a crucial aspect in the progression of the discipline of superconductivity, as it facilitates the expansion of the spectrum of feasible applications. Furthermore, through the investigation of non-traditional coupling mechanisms and the emergence of phenomena in high-temperature superconducting materials, scholars can acquire knowledge regarding fundamental principles of physics, including quantum significance, topological states, as well as systems of highly related electrons. The aforementioned observations not only augment our comprehension of superconductivity yet possess wider ramifications for other domains of physics as well as material research (Simutis et al., 2022; Kühn et al., 2006).

Furthermore, current research endeavors are focused on the advancement of fabrication methodologies and computational resources to facilitate the design and evaluation of high-temperature superconducting materials. The aforementioned technological progressions have the potential to expedite the identification and refinement of high-throughput screening (HS) substances, thereby enhancing the efficiency and cost-efficiency of the procedure (MacManus-Driscoll & Wimbush, 2021).

8.2. HISTORICAL BACKGROUND

The discovery of superconductivity was initially made by Heike Kamerlingh Onnes within the year 1911, while he was affiliated with the University of Leiden. It was observed by Onnes that the electrostatic attraction of mercury undergoes an abrupt transition to zero upon being subjected to temperatures beneath 4.2°K. This significant finding heralded the commencement of a novel epoch in the field of physics. Upon recognizing the phenomenon of superconductivity, scholars expeditiously discerned its potential to catalyze groundbreaking technological progressions through the facilitation of electricity transfer and the development of potent magnets (Gu, 2003; Flores-Livas et al., 2020; Felt & Nowotny, 1992).

8.2.1. Discovery of HTS Materials

For numerous decades, the primary emphasis of studies on superconductivity had been on traditional superconductors, which necessitated exceedingly low temperatures in proximity to absolute zero to manifest superconductivity. In the year 1986, a noteworthy advancement was made with the identification of materials exhibiting high-temperature superconductivity (HTS). The phenomenon of superconductivity found in a ceramics compound comprising copper, lanthanum, as well as oxygen was discovered by Georg Bednorz along with Alex Müller during their tenure at IBM. The compound, denoted as LBCO or just YBa2Cu3O7 manifested the phenomenon of superconductive properties at temperatures surpassing its boiling point for liquid nitrogen, which is 77°K. The finding was significant as it enabled the utilization of a comparatively affordable and readily available coolant, nitrogen in its liquid form, in lieu of the costly and intricate liquid helium that is mandatory for traditional superconductors (Jansen, 1995; Sato, 2008).

8.2.2. Milestones in HTS Research

The identification of high-temperature superconducting (HTS) materials has instigated a proliferation of scholarly investigation and has resulted in noteworthy achievements in the discipline. Other families of high-temperature superconducting (HTS) materials were expeditiously identified by researchers, after the discovery of LBCO. The compounds encompassed those based on bismuth, thallium, as well as mercury. The families demonstrated elevated critical temperatures, thereby extending the frontiers of high-temperature superconductivity (Masur et al., 2002) (Figure 8.1).

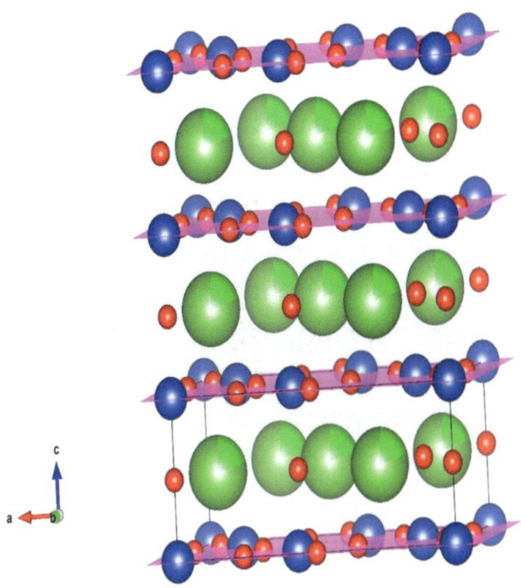

Figure 8.1. *Image of nano-sheet of LBCO.*

Source: Crystallography, Creative Commons License.

Remember

Navigating the challenges and capitalizing on the opportunities for integrating and commercializing HTS devices is crucial for advancing their practical applications across various fields.

The year 2008 marked a significant achievement in the field of superconductivity with the identification of superconductors based on iron. The present study highlights the superconducting properties of iron-based compounds, commonly referred to as pnictides or just chalcogenides, which have been observed to exhibit superconductive properties at temperatures exceeding 50°K. The identification of iron-based superconducting materials has broadened the scope of accessible materials for scientific inquiry and has introduced novel prospects for comprehending the underlying mechanisms of high-temperature superconductivity. To gain a comprehensive understanding of HTS, scholars have undertaken investigations into non-traditional pairing processes that deviate from the established Bardeen-Cooper-Schrieffer (BCS) hypothesis. The researchers investigated the function of robust electron associations, spin oscillations, and the interrelationship between superconductive properties and magnetism within high-temperature superconducting (HTS) materials. The objective of this investigation was to elucidate the underlying mechanisms that give rise to superconductive properties at high temperatures (Bruzzone et al., 2018; Werfel et al., 2006).

The progress made in the methods of synthesis along with fabrication has been a crucial factor in the research of high-temperature superconductors. The utilization of thin-film deposition techniques, including the use of pulsed lasers as well as molecular light epitaxy, has facilitated the production of superior-quality films along with heterogeneous structures that exhibit improved superconducting characteristics. The utilization of these methodologies has facilitated the assimilation of HS materials into diverse devices along with applications. Furthermore, materials exhibiting the phenomenon of HTS have been utilized in various practical applications across multiple disciplines. Effective electricity transmission has been made possible through the development of high-temperature superconductive wires, which effectively minimize energy losses within electrical grids. High-temperature superconducting (HTS) magnets have found applications in various fields such as MRI (magnetic resonance imaging) mechanisms, particle accelerators, as well as fusion research due to their exceptional performance. The applications have demonstrated the potential influence of HS on diverse sectors and have stimulated additional investigation and advancement (Mehta, 2011; Driscoll et al., 2000, Yamada et al., 2005).

8.3. FUNDAMENTAL MECHANISMS OF HTS

8.3.1. Understanding Conventional Superconductivity

The BCS hypothesis (Bardeen-Cooper-Schrieffer hypothesis) is the basis of our comprehension of superconducting acts, specifically traditional superconductivity. In the realm of conventional superconductivity, Cooper pairs have been created by electrons through the swap of lattice resonances or phonons. The aesthetically pleasing interplay between particles results in the emergence of a superconductive phase that is distinguished by the absence of voltage resistance as well as the repulsion of magnetic fields, commonly referred to by the term Meissner phenomenon (Kaur et al., 2021).

The BCS theory has demonstrated its efficacy in elucidating several crucial phenomena in traditional superconductors, such as the energy deficit, which denotes an assortment of energies devoid of electron states, as well as the threshold temperature (Tc) that marks the onset of superconductivity. The theoretical framework elucidates the process by which the shift from the conventional conducting phase to the superconducting phase occurs, which is attributed to the condensing of Cooper groups of two (Bozovic & Ahn, 2014; Gawith et al., 2019; Wu et al., 2022) (Figure 8.2).

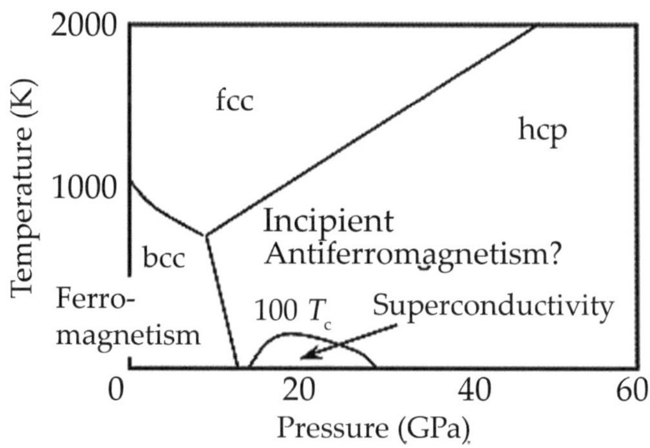

Figure 8.2. *Graphical representation of BCS theory.*

Source: Jozef Spaek, Creative Commons License.

8.3.2. Challenges in Explaining HTS Mechanisms

The phenomenon of (HTS) presents notable obstacles that diverge from the established conventional comprehension. A significant obstacle pertains to the detection of superconductivity at notably elevated temperatures in comparison to traditional superconductors, surpassing the projections of BCS theory. One of the primary objectives in research on HTS is to comprehend the underlying mechanisms that contribute to the heightened critical temperature (Meyer et al., 1997; Fan et al., 2016).

Moreover, materials with HTS display atypical characteristics and intricate phase representations. The scientific comprehension of HTS mechanisms is complicated by the interplay of computerized, lattice, as well as magnetic characteristics in these materials. The usual electron-phonon interactions, as posited by the BCS hypothesis, seem inadequate in accounting for the distinctive features of HTS (Nel et al., 2013; Avouris & Walkup, 1989).

8.3.3. Promising Theories and Models for HTS

In response to the challenges presented by HTS, different hypotheses and models were put forth, each providing distinct perspectives on how it works. According to a significant hypothesis, it is suggested that HTS content exhibits an unorthodox pairing uniformity referred to as d-wave uniformity, in contrast to the s-wave balance that is observed in typical superconductors. The manifestation of anisotropic behavior and the elucidation of certain empirical findings, including the angular reliance of the gaps in energy, are able to be attributed to the d-wave uniformity (Kunz-Schughart et al., 2004; Ischia & Fiori, 2021).

Furthermore, it is widely acknowledged that high-temperature superconducting (HTS) materials exhibit characteristics of highly correlated electron structures, wherein the interactions between electrons are of paramount importance. In comparison to traditional superconductors that are commonly characterized as Fermi fluids with weak interactions, high-temperature superconducting (HTS) materials display properties that are linked to significant relationships. The aforementioned interactions have the potential to give rise to unprecedented phases and phenomena within the realm of high-temperature superconductivity (Meyer & Kieras, 1997).

The scientific community has conducted thorough research on the significance of spin shifts as well as magnetic stimulation in high-temperature superconductors (HTS). The prevalent notion is that the presence of spin shifts can augment the phenomenon of superconductivity by acting as a mediator for the attractive interactions among electrons. Comprehending the correlation between spin behavior as well as superconductivity is of utmost importance in deciphering the mechanisms of high-temperature superconductivity. The investigation of high-temperature superconductivity has garnered interest with regard to the closeness to quantum thresholds and the phenomenon of quantum significance. Quantum criticality pertains to the characteristics exhibited by a substance in close proximity to a phase shift occurring at the temperature of absolute zero. The proposition posits that closeness to quantum thresholds may augment superconductivity in high-temperature superconducting materials and contribute to the determination of their distinctive characteristics (Cimperman et al., 2016).

Moreover, materials with a HTS display intriguing phenomena, including the competition among various order variables along with the manifestation of phase separation. The intricate behaviors exhibited by high-temperature superconductors pose a challenge to conventional comprehension of superconductivity, necessitating the formulation of novel theoretical structures to account for these mechanisms.

In recent times, there has been a utilization of high-energy physics methodologies to explore the properties of high-temperature superconductors (HTS). The utilization of gauge hypotheses, string hypotheses, as well as the AdS/CFT correspondence has yielded alternative viewpoints regarding the essence of superconductivity. The utilization of concepts derived from the study of particles is employed to explicate the conduct of high-temperature superconducting (HTS) materials, thereby providing novel perspectives into their fundamental mechanisms. Furthermore, the domain of topological superconductive properties has garnered considerable interest in the realm of high-temperature superconductivity investigation. Materials that demonstrate exceptional characteristics, such as the existence of Majorana fermions, which are particles that possess the qualities of both a particle and its antiparticle, are referred to as topological superconducting materials The identification of Majorana fermions within high-temperature superconducting materials has presented novel prospects for both essential scientific inquiry and potential utilization in the realm of quantum information processing (Nel et al., 2013).

Models and hypotheses in unison offer a comprehensive structure to comprehend (HTS) and steer continuous research endeavors toward unraveling the underlying mechanisms that facilitate it. Further investigation of such hypotheses, in conjunction with experimental progress, is imperative in order to fully realize the capabilities of HTS and propel its implementation in diverse disciplines (Falco et al., 2011).

8.4. NOVEL HTS MATERIALS

The phenomenon of (HTS) has become the focus of extensive scholarly inquiry, resulting in the identification of various material groups that have gained significant recognition in this domain. Cuprite superconductors, which were the initial superconductors with high temperatures to be identified, have attracted considerable interest. The compounds exhibit stratified configurations comprising copper oxide planes. Their distinctive crystal structures along with electronic characteristics are instrumental in conferring upon them their remarkable superconducting attributes. Cuprite superconductors are known to demonstrate relatively elevated critical temperatures (Tc) that can attain a maximum of 138°K (–135°C). This attribute renders them significant in comprehending the mechanisms that underlie HTS. The category of high-temperature superconducting (HTS) materials includes a significant group of iron-based superconducting materials The compounds are comprised of coats of metallic pnictides or just chalcogenides, while they manifest unique crystal structures as well as electronic characteristics in contrast to cuprates. The electronic structure of iron-based superconductors has been defined by multiple bands and significant electron-electron correlations, resulting in non-traditional pairing mechanisms. Scholars have directed their attention toward comprehending these substances and investigating their capacity to attain elevated critical temperatures (Hayakawa et al., 2016; Liu et al., 2023).

Extensive research has been conducted on heavy fermion superconducting materials which are distinguished by the inclusion of heavy elements such as rare earth or just actinide compounds. The materials demonstrate robust electron-electron relationships and atypical pairing mechanisms, which offer valuable perspectives on the interdependence of potent electronic correlations as well as superconductivity. Significant fermion superconducting materials demonstrate a diverse spectrum of extreme temperatures, frequently inferior to those of cuprates as well as iron-based materials. Nevertheless, they are significant in comprehending the fundamental physics of high-temperature superconductivity (Zhuravel et al., 2006; Mohanty et al., 2004).

Apart from the material groups, organic superconductors are attracting significant interest owing to their distinctive characteristics and phenomena. The compounds are composed of organic molecules that are organized within crystal lattices as well as demonstrate the phenomenon of superconductive behavior at temperatures that are comparatively low. The intrinsic characteristics of these substances provide the potential for customization and adaptability, rendering them captivating for exploring unorthodox superconductivity. The investigation of organic superconductors enhances our comprehension of the function of molecular bonds in high-temperature superconductivity phenomena (Liu et al., 2019; Kopaev et al., 2013; Sunwong et al., 2013; Jin et al., 2016; Grilli et al., 2013).

8.4.1. Properties and Characteristics of Unconventional HTS Materials

Unconventional high-temperature superconducting (HTS) materials exhibit a notable characteristics in the form of a comparatively elevated critical temperature (Tc) in contrast with typical superconductors. It is noteworthy that conventional superconductors necessitate exceedingly low temperatures within the vicinity of absolute zero. However, high-temperature superconducting (HTS) materials have the potential to manifest superconductive properties at temperatures surpassing the boiling limit for liquid nitrogen (-196°C). The aforementioned attribute endows high-temperature superconducting materials with enhanced practicality across diverse applications and augments the scope for technological progressions. Scholars persist in investigating and expanding the limits of critical temperature values in order to attain elevated temperatures. Materials with unconventional high-temperature superconductivity display pairing processes that differ from those predicted by the conventional Bardeen-Cooper-Schrieffer (BCS) theory. The promotion of superconductivity at elevated temperatures is significantly influenced by various factors such as strong electron-electron relationships along with spin fluctuations. Comprehending these atypical pairing mechanisms is imperative for the progression of our comprehension of high-temperature superconducting (HTS) materials and the establishment of novel theoretical structures that can elucidate their conduct (Varlamov et al., 1999; Gross et al., 1997; Rahman et al., 2015) (Figure 8.3).

Anisotropy is a salient feature that is frequently observed in a multitude of high-temperature superconducting (HTS) stuff. Anisotropy is a phenomenon characterized by the variability of properties in relation to distinct crystallographic orientations. Materials with high-temperature superconductivity can display varying conductivities, magnetic reactions, as well as critical current concentrations in distinct axes. The behavior of high-temperature superconducting (HTS) materials is influenced by the coherency length, which is a parameter that signifies the spatial distribution of superconductive electron pairs. The characteristics are imperative in the process of enhancing the performance of HS materials for particular applications and devising effective superconductive apparatus (Romano et al., 2018; Fischer et al., 2007; Schiavoni et al., 2016).

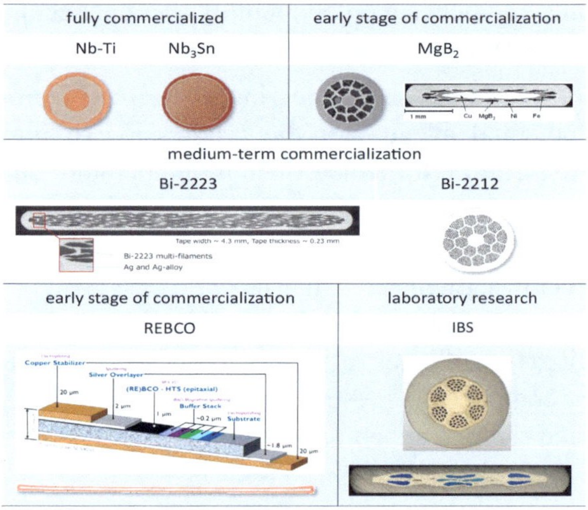

Figure 8.3. *Images of superconducting materials.*

Source: Chao Yao & Yanwei Ma, Creative Commons License.

The phenomena of flux pinning as well as critical voltage density (Jc) are of notable importance in the realm of high-temperature superconducting (HTS) materials. The phenomenon of flux pinning pertains to the capacity of a substance to confine magnetic field lines, thereby exerting an impact on the superconductive characteristics of the substance. The crucial current density is defined as the upper limit of current density that a superconductor may withstand while maintaining its superconducting properties. Comprehending the concept of flux pinning as well as Jc holds paramount importance in the development of high-performance along with stable HTS-based electronics (Rokhvarger et al., 2003; Chen et al., 2004; Gökalp et al., 2018).

8.4.2. Recent Advances in Discovering New HTS Materials

In recent times, scholars have made noteworthy advancements in the exploration of novel HTS materials, thereby broadening the scope of elevated temperatures superconductivity. The implementation of crystal engineering along with doping techniques has been instrumental in the advancement of innovative high-temperature superconducting materials. Scientists have successfully manipulated crystalline structures as well as formulations of materials to improve their superconducting characteristics. The utilization of doping techniques, which involve the introduction of particles or atoms into the structure of a crystal, has been implemented to alter the electronic characteristics and augment superconductivity (Gardner et al., 2004; Noh et al., 2020; Gubser, 2004).

The utilization of interface engineering as well as heterostructures has surfaced as a potential avenue for the progression of high-temperature superconductivity. The performance of high-temperature superconducting (HTS) materials can be improved by researchers through the creation of artificial connections between various materials or the construction of multilayered structures. Engineered interfaces possess distinct properties that facilitate elevated critical temperatures as well as enhanced superconducting features. The utilization of heterostructures, which entail the arrangement of diverse layers of components, presents prospects for customizing electronic characteristics and refining superconductivity (Mayr & Bojanic, 2009; Chu et al., 2015).

The field of HTS has been expanded by advancements in nanotechnology along with defect engineering, providing novel opportunities for research. Manipulation of HTS behavior has been achieved by utilizing nanoscale characteristics and defects, as demonstrated by researchers. Through the manipulation of substances at the nanoscale, researchers have successfully enhanced the superconducting characteristics and fabrication methodologies of nanostructured high-temperature superconducting materials (Jin, 2017).

Moreover, recent studies have emphasized the investigation of non-traditional high-temperature superconducting (HTS) materials. The exploration of materials that include topological superconductors, which exhibit distinctive electronic characteristics and unusual phenomena, is being undertaken to assess their possibilities for employment in elevated temperatures superconductivity. The investigations broaden the scope of HS and present novel opportunities for comprehending and utilizing superconductivity in various substances. Through the analysis of diverse high-temperature superconducting material groups, their respective attributes, and atypical features, scholars can acquire a deeper comprehension of the fundamental mechanisms governing elevated temperatures superconductivity. The latest developments in the identification of novel high-temperature superconducting (HTS) materials have presented promising opportunities for expanding the limits of superconductive technologies and investigating unexplored domains within this domain (Zhou et al., 2020; Villamor et al., 2019; Gao et al., 2021).

8.5. HIGH-TEMPERATURE SUPERCONDUCTING DEVICES

An overview of the applications of HTS devices. The distinctive characteristics and capabilities of elevated temperatures superconducting equipment have the ability to bring about a revolutionary impact across diverse industries and sectors. High-temperature superconducting (HTS) materials exhibit the remarkable property of zero electrical resistance when subjected to temperatures exceeding the melting point for liquid nitrogen, approximately 77° or just –196°C. The aforementioned attribute renders HTS devices exceedingly appealing for a diverse array of applications that necessitate superior efficacy, reduced dimensions, and enhanced functionality (Newman & Lyons, 1993).

8.5.1. Current State of HTS Devices in Various Fields

1. **The Energy Sector:** High-temperature superconducting (HTS) devices present notable benefits within the energy industry, particularly in the areas of augmenting efficiency and mitigating energy losses. High-temperature superconductor (HTS) power cables have the capability to transmit substantial quantities of electricity over extended distances with negligible losses, thereby enhancing the effectiveness of the electricity grid on the whole. The application of high-temperature superconducting (HTS) materials in transformers as well as fault current constraints is known to exploit their distinctive properties, leading to the development of compact and effective devices that augment the predictability and consistency of electricity systems. Moreover, HTS generators as well as motors exhibit superior power densities and enhanced efficiency in contrast to traditional technologies, rendering them highly appropriate for utilization in energy from renewable sources production and electric transportation (Hull et al., 2004).

2. **Transportation High-Temperature Superconducting (HTS):** These devices possess the capability to transform transportation systems through the facilitation of increased power densities, enhanced efficiency, and decreased emissions. High-temperature superconducting (HTS) materials have the potential to significantly enhance the performance of maglev trains, which rely on strong magnetism to accomplish both levitation and propulsion. High-temperature superconducting (HTS) magnets have the capability to produce the requisite magnetic fields, thereby facilitating the creation of high-velocity magnetic levitation (maglev) trains that exhibit decreased energy consumption and frictional forces. The utilization of high-temperature superconducting (HTS) springs in the superconductive magnetic battery (SMES) systems present a promising solution for storing electrical energy within magnetic fields, thereby enabling efficient energy retention and retrieval in diverse transportation applications. Moreover, the utilization of HTS-based power electronics has the potential to substantially augment the efficacy and functionality of electrically powered cars, ships, as well as aircraft, thereby promoting a more environmentally conscious and sophisticated transportation system (Wimbush & Strickland, 2016).
3. **Medicine High-Throughput Sequencing (HTS):** These devices exhibit potential in the medical domain, specifically in the areas of imaging and diagnostics. High-temperature superconducting (HTS) magnets are essential components in MRI (magnetic resonance imaging) technology, facilitating the production of high magnetic fields that are necessary for the acquisition of precise and intricate medical images. The utilization of high-temperature superconducting (HTS) magnets in magnetic resonance imaging (MRI) apparatuses results in an enhancement of the capacity for imaging as well as an improvement in diagnostic precision. The utilization of high-temperature superconducting (HTS) sensors within magnetoencephalography (MEG) has proven advantageous in accurately detecting the magnetic radiation generated within the brain of a person during neural activity mapping. High-temperature superconducting (HTS) contents have potential applications within specific systems for drug delivery. By utilizing superconductive magnets, drug-carrying tiny particles are able to be directed to exact locations within the

body, allowing for controlled and accurate drug administration (Grilli, 2016; Hull, 2003).

8.5.2. Challenges and Opportunities for HTS Device Integration and Commercialization

Although HTS devices hold significant promise, there exist a number of obstacles that must be overcome in order to achieve their effective integration as well as commercialization.

1. The topic of interest pertains to the domains of manufacturing along with cost. The challenges of paramount importance in the realm of high-temperature superconducting (HTS) devices pertain to the growing production processes and the concomitant reduction of manufacturing costs. At present, the cost of HTS materials is comparatively high, necessitating the development of large-scale manufacturing methodologies to reduce the total expense of devices. The current study and development endeavors are centered on enhancing the efficiency of production procedures and investigating substitute techniques of manufacturing to render high-temperature superconducting devices more financially feasible (Nisenoff, 1995).
2. The study focuses on the properties and efficacy of materials used in a given context. Improving the efficiency of high-temperature superconducting (HTS) materials is imperative for wider acceptance and utilization of HTS devices. The enhancement of critical temperatures (Tc) along with crucial current densities (Jc) is a persistent research objective. Elevating the critical temperature (Tc) of high-temperature superconducting (HTS) devices would enhance their operational capacity at elevated temperatures, thereby rendering them more convenient and manageable. Improving the critical current density (Jc) is a crucial aspect in the development of high-temperature superconducting (HTS) devices that are both efficient and compact, capable of carrying increased current densities (Marchionini et al., 2017).
3. The process of combining different subsystems or components into a single system is referred to as system integration. The incorporation of HS devices into pre-existing systems and structures presents various technical hurdles. The development of effective interfaces, interfaces, as well as cooling structures that

are capable of withstanding the rigorous operational parameters of high-temperature superconducting (HTS) devices is a crucial domain of investigation. Effective thermal management is of paramount importance for preserving the superconducting properties of HTS materials. To address this issue, scholars are investigating novel cooling methodologies as well as materials to surmount these obstacles (Soomro et al., 2020).

4. The topics of standardization along with certification are of great importance. The development of norms and procedures for accreditation is crucial for the broad adoption and marketing of HTS electronics. The task at hand pertains to the formulation of standards for the operational efficiency, dependability, security, and compatibility of devices in diverse sectors and domains. The implementation of standardization measures is expected to enhance the reliability of HS technologies along with streamlining their assimilation into established frameworks.
5. Collaboration along with funding are two important factors that contribute to the success of a project. The advancement of HTS studies and experiments seems contingent upon the crucial collaboration among academia, industry, as well as government agencies. Collaboration among individuals and groups facilitates the exchange of information, sharing of materials, and implementation of multidisciplinary strategies, thereby expediting the pace of innovation along with technological progress. It is imperative to provide additional financial resources and assistance towards the development of HTS devices in order to facilitate significant advancements, surmount obstacles, and expedite the process of bringing these technologies to market (Malozemoff, 2012).

8.6. EMERGENT RESEARCH DIRECTIONS QUANTUM MATERIALS AND HTS

8.6.1. Exploration of Quantum Criticality in HTS Materials

Scholars are investigating the domain of quantum criticality in order to comprehend its impact on materials exhibiting elevated temperatures of superconductive properties. Quantum criticality pertains to the conduct of a substance in proximity to a quantum phase switch, wherein minute alterations in extrinsic circumstances can result in substantial modifications in its characteristics. The objective of researchers is to comprehend the complex correlation between quantum importance and the development and inhibition of superconductivity by examining HTS contents in proximity to these crucial locations. The researchers conduct an investigation into the impact of various factors such as magnetic fields, force, along with doping upon the quantum crucial behavior of high-temperature superconducting (HTS) materials. This study provides insights into the fundamental mechanisms that drive HTS phenomena (Aynajian et al., 2012).

The significance of emergent phases along with topological states in HTS is a topic of interest in academic research. Researchers are currently directing their attention towards comprehending the function of emergent phases along with topological states in high-temperature superconducting (HTS) materials, alongside their investigation of quantum criticality. These materials frequently demonstrate unique electronic phases, including charge density vibrations or nematicity, which may exist alongside or contend with superconductivity. The objective of scientists is to clarify the influence of emergent stages on the actions and features of high-temperature superconducting materials through investigation. In addition, the researchers are currently engaged in the examination of the feasibility of topological superconductive properties, an exceptional form of matter that provides a foundation for Majorana fermions. The comprehension and utilization of emergent phases along with topological states have the potential to facilitate creativity in quantum information processing alongside more sophisticated technologies (Bauer et al., 2020).

8.6.2. Novel Fabrication Techniques for HTS Materials

8.6.2.1. Advances in Thin-Film Growth and Synthesis

Considerable advancements have been achieved in the development of sophisticated methods for the development and synthesis of thin films pertaining to high-temperature superconducting materials. Researchers are presently investigating novel techniques for manufacturing high-quality thin sheets of HTS substances, which allow for meticulous regulation of their setup, structure, as well as interface properties. Through the utilization of epitaxial growth methodologies, scholars are capable of manipulating the arrangement of crystals in order to improve the efficacy of High-Temperature Superconducting (HTS) lightweight films. The researchers are currently examining the impact of strain technology and substrate selection regarding the superconducting characteristics of said films. The progress made in the development along with the synthesis of lightweight films has the capacity to augment the efficacy and durability of high-temperature superconducting (HTS) devices (Persky et al., 2022).

Tip

Staying updated on novel fabrication techniques for high-temperature superconducting (HTS) materials is essential for pushing the boundaries of their performance and applicability.

8.6.2.2. Nanostructuring and Interface Engineering for Enhanced Performance

The enhancement of performance in high-temperature superconducting (HTS) materials is a subject of significant research interest, with a particular focus on nanostructuring as well as interface technology. Nanoscale manufacturing methods are being utilized by scientists to produce artificial structures as well as heterogeneous structures within contents exhibiting HTS. The manipulation of nanostructures can exert an effect on crucial parameters such as crucial current density as well as vortex pinning, among other superconductive properties. This presents a promising avenue for improving the efficiency of high-temperature superconductors. Furthermore, scholars are investigating the potential of interface technology, whereby meticulously crafted interfaces between HS materials alongside different compounds may result in enhanced device properties and durability. The novel methodologies exhibit the potential for expanding the limits of HS technology (Lee et al., 2021).

8.6.3. Computational Modeling and Simulations

8.6.3.1. Quantum Monte Carlo Methods for HTS Predictions

The utilization of computational models and simulations is of paramount importance in comprehending the intricate behavior of high-temperature superconducting (HTS) substances. Quantum Monte Carlo (QMC) modeling is a potent methodology that facilitates the investigation of the characteristics of electron systems exhibiting strong correlations, such as high-temperature superconducting (HTS) materials. Quantum Monte Carlo (QMC) techniques have the potential to offer valuable insights into the determination of crucial temperatures, phase schematics, as well as transportation properties, thereby facilitating a deeper comprehension and more accurate forecasting of HTS behavior. Through the simulation of the interaction between superconductivity as well as magnetism, researchers can investigate the complex correlation between these occurrences and acquire a more profound understanding of the fundamental mechanisms that propel high-temperature superconductivity (Burns et al., 2019) (Figure 8.4).

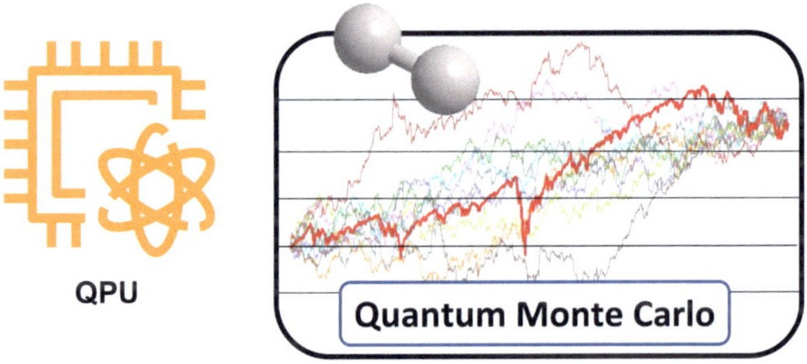

Figure 8.4. *Illustration of quantum Monte Carlo on quantum computers.*

Source: AWS, Creative Commons License.

Did You Know?

Emerging research in high-temperature superconductivity is focused on exploring unconventional pairing mechanisms that go beyond traditional theories, offering new insights into the fundamental behavior of superconductors.

8.6.3.2. Machine Learning and AI Approaches for HTS Material Discovery

The utilization of machine learning (ML) along with artificial intelligence (AI) has become increasingly significant in expediting the identification of novel HS materials. Scholars are currently engaged in the development of algorithms and theories that have the capacity to forecast superconducting properties by taking into account the chemical makeup and physical makeup of materials. By employing high-throughput computing screening techniques on extensive material databases, machine learning, and artificial intelligence methodologies can effectively recognize potential high-throughput candidate selections that exhibit favorable attributes. The utilization of this computational methodology results in a notable reduction in both the temporal and financial resources required for conventional trial-and-error experimentation. This, in turn, creates novel opportunities for expeditious exploration and development of HS materials (Imtiaz et al., 2014).

8.6.4. Unconventional Pairing Mechanisms

8.6.4.1. Role of Strong Electron Correlations and Spin Fluctuations

The investigation of non-traditional pairing mechanisms represents a crucial avenue of inquiry in the field of high-temperature superconductivity. The scientific community is currently conducting research on the impact of robust electron-electron relationships as well as spin fluctuations on the facilitation or impediment of superconductive properties in high-temperature superconducting (HTS) materials. Unusual pairing processes frequently entail intricate interplay among various levels of freedom, including charge, spin, as well as lattice dynamics. Researchers seek to elucidate the fundamental mechanisms that underlie the observable high-temperature superconductive properties in those substances through the investigation of electron correlations along with spin fluctuations (Scheurer et al., 2015).

8.6.4.2. Majorana Fermions and Topological Superconductivity

The subject of Majorana fermions along with topological superconductivity has garnered considerable interest within the realm of high-temperature superconductivity study. Researchers are currently investigating the feasibility of achieving Majorana fermions, which are unique particles that have the potential to serve as a crucial component in topological quantum computation, through the utilization of high-temperature superconducting (HTS) materials. Researchers endeavor to generate and identify Majorana fermions within high-temperature superconductor (HTS) systems by meticulously designing circumstances, such as nearness to topological barriers or utilization of unorthodox pairing states. The investigation of topological superconductive properties and the examination of these distinct particles possess significant prospects for promoting both fundamental physics as well as technological implementations (Inosov et al., 2011; Curro et al., 2005).

The emergent areas of study within elevated temperature superconductivity are interdisciplinary in nature, spanning various scientific disciplines such as condensation physics, the study of materials, as well as computational simulation. Researchers endeavor to unlock all the possibilities of elevated temperatures superconductivity as well as pave the path for transformative advances in technology by enhancing their comprehension of quantum materials, producing innovative HTS structures, utilizing advanced computational techniques, and exploring unconventional pairing processes.

Activity 1.1

You are leading a brainstorming session with a team of scientists and researchers focused on high-temperature superconductivity. Encourage them to propose innovative research directions or experiments that could potentially push the boundaries of current understanding and pave the way for exciting breakthroughs.

8.7. SUMMARY

Chapter eight delves into the emerging research and innovative directions in HTS. It begins with an introduction to the chapter's topics and provides a historical background on the discovery of HTS materials and significant milestones in HTS research. The chapter explores the fundamental mechanisms of HTS, including the challenges in explaining these mechanisms and promising theories and models for HTS. It discusses novel HTS materials, their properties, characteristics, and recent advances in discovering new HTS materials. Furthermore, the chapter examines the current state of HTS devices in various fields, the challenges and opportunities for their integration and commercialization. Additionally, it explores the emergent research directions in quantum materials and HTS, such as the exploration of quantum criticality in HTS materials, novel fabrication techniques, computational modeling and simulations, and unconventional pairing mechanisms.

REVIEW QUESTIONS

1. Discuss the historical background of high-temperature superconductivity, including the discovery of HTS materials and milestones in HTS research.
2. Explain the challenges in understanding the fundamental mechanisms of HTS.
3. What are some promising theories and models for explaining HTS behavior?

4. Describe the properties and characteristics of unconventional HTS materials.
5. What are recent advances in the discovery of new HTS materials?
6. Discuss the current state of HTS devices in various fields and their applications.
7. What are the challenges and opportunities for integrating and commercializing HTS devices?
8. How does the exploration of quantum criticality in HTS materials contribute to emergent research directions?
9. What are some novel fabrication techniques for HTS materials?

REFERENCES

1. Avouris, P., & Walkup, R. E. (1989). Fundamental mechanisms of desorption and fragmentation induced by electronic transitions at surfaces. *Annual Review of Physical Chemistry, 40*(1), 173–206.
2. Aynajian, P., da Silva Neto, E. H., Gyenis, A., Baumbach, R. E., Thompson, J. D., Fisk, Z., & Yazdani, A. (2012). Visualizing heavy fermions emerging in a quantum critical Kondo lattice. *Nature, 486*(7402), 201–206.
3. Bauer, B., Bravyi, S., Motta, M., & Chan, G. K. L. (2020). Quantum algorithms for quantum chemistry and quantum materials science. *Chemical Reviews, 120*(22), 12685–12717.
4. Bozovic, I., & Ahn, C. (2014). A new frontier for superconductivity. *Nature Physics, 10*(12), 892–895.
5. Bruzzone, P., Fietz, W. H., Minervini, J. V., Novikov, M., Yanagi, N., Zhai, Y., & Zheng, J. (2018). High temperature superconductors for fusion magnets. *Nuclear Fusion, 58*(10), 103001.
6. Burns, S. R., Sando, D., Xu, B., Dupé, B., Russell, L., Deng, G., & Ulrich, C. (2019). Expansion of the spin cycloid in multiferroic BiFeO3 thin films. *NPJ Quantum Materials, 4*(1), 18.
7. Chen, L. F., Ong, C. K., Neo, C. P., Varadan, V. V., & Varadan, V. K. (2004). *Microwave Electronics: Measurement and Materials Characterization.* John Wiley & Sons.
8. Chu, C. W., Deng, L. Z., & Lv, B. (2015). Hole-doped cuprate high temperature superconductors. *Physica C: Superconductivity and Its Applications, 514*, 290–313.
9. Cimperman, M., Brenčič, M. M., & Trkman, P. (2016). Analyzing older users' home telehealth services acceptance behavior—applying an Extended UTAUT model. *International Journal of Medical Informatics, 90*, 22–31.
10. Curro, N. J., Caldwell, T., Bauer, E. D., Morales, L. A., Graf, M. J., Bang, Y., & Sarrao, J. L. (2005). Unconventional superconductivity in PuCoGa5. *Nature, 434*(7033), 622–625.
11. Driscoll, D., Dombrovski, V., & Zhang, B. (2000). Development status of superconducting motors. *IEEE Power Engineering Review, 20*(5), 12–15.
12. Falco, C., Baccile, N., & Titirici, M. M. (2011). Morphological and structural differences between glucose, cellulose and lignocellulosic biomass derived hydrothermal carbons. *Green Chemistry, 13*(11), 3273–3281.
13. Fan, J., Wang, Y., Rosenfeld, D., & Liu, X. (2016). Review of aerosol–

cloud interactions: Mechanisms, significance, and challenges. Journal of the Atmospheric *Sciences, 73*(11), 4221–4252.

14. Felt, U., & Nowotny, H. (1992). Striking gold in the 1990s: the discovery of high-temperature superconductivity and its impact on the science system. *Science, Technology, & Human Values, 17*(4), 506–531.
15. Fischer, Ø., Kugler, M., Maggio-Aprile, I., Berthod, C., & Renner, C. (2007). Scanning tunneling spectroscopy of high-temperature superconductors. *Reviews of Modern Physics, 79*(1), 353.
16. Flores-Livas, J. A., Boeri, L., Sanna, A., Profeta, G., Arita, R., & Eremets, M. (2020). A perspective on conventional high-temperature superconductors at high pressure: Methods and materials. *Physics Reports, 856*, 1–78.
17. Gao, K., Shaabani, S., Xu, R., Zarganes-Tzitzikas, T., Gao, L., Ahmadianmoghaddam, M., & Dömling, A. (2021). Nanoscale, automated, high throughput synthesis and screening for the accelerated discovery of protein modifiers. *RSC Medicinal Chemistry, 12*(5), 809–818.
18. Gardner, C. R., Walsh, C. T., & Almarsson, Ö. (2004). Drugs as materials: valuing physical form in drug discovery. *Nature Reviews Drug Discovery, 3*(11), 926–934.
19. Gawith, J., Geng, J., Ma, J., Shen, B., Li, C., & Coombs, T. A. (2019). HTS transformer–rectifier flux pump optimization. *IEEE Transactions on Applied Superconductivity, 29*(5), 1–5.
20. Gökalp, İ., Uz, V. E., Saltan, M., & Tutumluer, E. (2018). Technical and environmental evaluation of metallurgical slags as aggregate for sustainable pavement layer applications. *Transportation Geotechnics, 14*, 61–69.
21. Grilli, F. (2016). Numerical modeling of HTS applications. *IEEE Transactions on Applied Superconductivity, 26*(3), 1–8.
22. Grilli, F., Pardo, E., Stenvall, A., Nguyen, D. N., Yuan, W., & Gömöry, F. (2013). Computation of losses in HTS under the action of varying magnetic fields and currents. *IEEE Transactions on Applied Superconductivity, 24*(1), 78–110.
23. Gross, R., Alff, L., Beck, A., Froehlich, O. M., Koelle, D., & Marx, A. (1997). Physics and technology of high temperature superconducting Josephson junctions. *IEEE Transactions on Applied Superconductivity, 7*(2), 2929–2935.
24. Gu, C. (2023). The co-production of normal science: A social history of high-temperature superconductivity research in China (1987–2008). *Social Studies of Science, 53*(1), 81–101.

25. Gubser, D. U. (2004). Superconductivity: an emerging power-dense energy-efficient technology. *IEEE Transactions on Applied Superconductivity, 14*(4), 2037–2046.
26. Hayakawa, N., Ishida, K., Matsuoka, T., & Kojima, H. (2016). Electrical insulation characteristics and mechanisms of low dielectric loss materials for high-voltage HTS cables. *IEEE Transactions on Applied Superconductivity, 27*(4), 1–4.
27. Hull, J. R. (2003). Applications of high-temperature superconductors in power technology. *Reports on Progress in Physics, 66*(11), 1865.
28. Hull, J. R., & Murakami, M. (2004). Applications of bulk high-temperature superconductors. *Proceedings of the IEEE, 92*(10), 1705–1718.
29. Imtiaz, A., Wallis, T. M., & Kabos, P. (2014). Near-field scanning microwave microscopy: An emerging research tool for nanoscale metrology. *IEEE Microwave Magazine, 15*(1), 52–64.'
30. Inosov, D. S., Park, J. T., Charnukha, A., Li, Y., Boris, A. V., Keimer, B., & Hinkov, V. (2011). Crossover from weak to strong pairing in unconventional superconductors. *Physical Review B, 83*(21), 214520.
31. Ischia, G., & Fiori, L. (2021). Hydrothermal carbonization of organic waste and biomass: a review on process, reactor, and plant modeling. *Waste and Biomass Valorization, 12,* 2797–2824.
32. Jansen, D. (1995). Convergence of basic and applied research? Research orientations in German high-temperature superconductor research. Science, Technology, & Human Values, 20(2), 197–233.
33. Jin, C. Q. (2017). Using pressure effects to create new emergent materials by design. *MRS Advances, 2*(48), 2587–2596.
34. Jin, J. X., Tang, Y. J., Xiao, X. Y., Du, B. X., Wang, Q. L., Wang, J. H., & Zhu, J. G. (2016). HTS power devices and systems: principles, characteristics, performance, and efficiency. *IEEE Transactions on Applied Superconductivity, 26*(7), 1–26.
35. Kaur, H., Kaur, H., & Sharma, A. (2021). A review of recent advancement in superconductors. *Materials today: Proceedings, 37*, 3612–3614.
36. Kopaev, Y., & Pudalov, V. (2013). Introduction to the Special Issue—Selected Contributions to the 4th International Conference on Fundamental Problems of High Temperature Superconductivity (FPS'11). *Journal of Superconductivity and Novel Magnetism, 26,* 2807–2808.
37. Kühn, L., Müller, M., Schubert, R., Beyer, C., de Haas, O., & Schultz, L. (2006, September). Dynamic behavior of a linear superconducting levitation transport system using YBCO bulk material. In Proc.

Maglev, *Transactions on Applied Superconductivity*, *28*(8), 1-9.

38. Kunz-Schughart, L. A., Freyer, J. P., Hofstaedter, F., & Ebner, R. (2004). The use of 3-D cultures for high-throughput screening: the multicellular spheroid model. *Journal of biomolecular screening, 9*(4), 273–285.
39. Lee, K., Choe, J., Iaia, D., Li, J., Zhao, J., Shi, M., & Trivedi, N. (2021). Metal-to-insulator transition in Pt-doped TiSe2 driven by emergent network of narrow transport channels. *NPJ Quantum Materials, 6*(1), 8.
40. Liu, H., Ren, B., Guan, X., Wen, P., Zuo, T., Liu, H., & Zuo, T. (2019). Fundamental of HTS materials and microwave filter design. *High-Temperature Superconducting Microwave Circuits and Applications,1(2)* 15–37.
41. Liu, Y., Tian, X., Han, Y. C., Chen, Y., & Hu, W. (2023). High-temperature shock synthesis of high-entropy-alloy nanoparticles for catalysis. *Chinese Journal of Catalysis, 48*, 66–89.
42. MacManus-Driscoll, J. L., & Wimbush, S. C. (2021). Processing and application of high-temperature superconducting coated conductors. *Nature Reviews Materials, 6*(7), 587–604.
43. Malozemoff, A. P. (2012). Second-generation high-temperature superconductor wires for the electric power grid. *Annual Review of Materials Research, 42*, 373–397.
44. Marchionini, B. G., Yamada, Y., Martini, L., & Ohsaki, H. (2017). High-temperature superconductivity: a roadmap for electric power sector applications, 2015–2030. *IEEE Transactions on Applied Superconductivity, 27*(4), 1–7.
45. Masur, L. J., Kellers, J., Li, F., Fleshler, S., & Podtburg, E. R. (2002). Industrial high temperature superconductors: perspectives and milestones. *IEEE Transactions on Applied Superconductivity, 12*(1), 1145–1150.
46. Mayr, L. M., & Bojanic, D. (2009). Novel trends in high-throughput screening. *Current Opinion in Pharmacology, 9*(5), 580–588.
47. Mehta, S. (2011). US effort on HTS power transformers. *Physica C: Superconductivity and Its Applications, 471*(21–22), 1364–1366.
48. Meyer, D. E., & Kieras, D. E. (1997). A computational theory of executive cognitive processes and multiple-task performance: Part I. Basic mechanisms. *Psychological Review, 104*(1), 3.
49. Mohanty, P., Wei, J. Y. T., Ananth, V., Morales, P., & Skocpol, W. (2004). Nanoscale high-temperature superconductivity. *Physica C: Superconductivity, 408*, 666–669.
50. Nel, A., Xia, T., Meng, H., Wang, X., Lin, S., Ji, Z., & Zhang, H. (2013).

Nanomaterial toxicity testing in the 21st century: use of a predictive toxicological approach and high-throughput screening. *Accounts of Chemical Research, 46*(3), 607–621.

51. Newman, N., & Lyons, W. G. (1993). High-temperature superconducting microwave devices: Fundamental issues in materials, physics, and engineering. *Journal of Superconductivity, 6*, 119–160.
52. Nisenoff, M. (1995). Cryocoolers and High-Temperature Superconductors: Advancing toward Commercial Applications. *Cryocoolers 8,* 913–917.
53. Noh, J., Gu, G. H., Kim, S., & Jung, Y. (2020). Uncertainty-quantified hybrid machine learning/density functional theory high-throughput screening method for crystals. *Journal of Chemical Information and Modeling, 60*(4), 1996–2003.
54. Pamidi, S., Kim, C. H., & Graber, L. (2015). High-temperature superconducting (HTS) power cables cooled by helium gas. In *Superconductors in the Power Grid (pp. 225–260).* Woodhead Publishing.
55. Persky, E., Sochnikov, I., & Kalisky, B. (2022). Studying quantum materials with scanning SQUID microscopy. *Annual Review of Condensed Matter Physics, 13,* 385–405.
56. Rahman, M. A., Rahaman, M. Z., & Samsuddoha, M. N. (2015). A review on cuprate based superconducting materials including characteristics and applications. *American Journal of Physics and Applications, 3*(2), 39–56.
57. Rokhvarger, A. E., & Chigirinsky, L. A. (2003). Unconventional nanoparticle technology of superconductor ceramic articles. *MRS Online Proceedings Library, 776,* 5131–5136.
58. Romano, P., Riccio, M., Guarino, A., Martucciello, N., Grimaldi, G., Leo, A., & Nigro, A. (2018). Electron doped superconducting cuprates for photon detectors. *Measurement, 122*, 502–506.
59. Sato, K. I. (2008). Present status and future perspective of high-temperature superconductors. *SEI Technical Review: English Edition, 66*, 55.
60. Scheurer, M. S., & Schmalian, J. (2015). Topological superconductivity and unconventional pairing in oxide interfaces. *Nature Communications, 6*(1), 6005.
61. Schiavoni, S., Bianchi, F., & Asdrubali, F. (2016). Insulation materials for the building sector: A review and comparative analysis. *Renewable and Sustainable Energy Reviews, 62*, 988–1011.
62. Simutis, G., Küspert, J., Wang, Q., Choi, J., Bucher, D., Boehm, M., & Mazzone, D. G. (2022). Single-domain stripe order in a high-temperature superconductor. *Communications Physics, 5*(1), 296.

63. Soomro, W. A., Guo, Y., Lu, H. Y., & Jin, J. X. (2020, October). Advancements and impediments in applications of high-temperature superconducting material. In *2020 IEEE International Conference on Applied Superconductivity and Electromagnetic Devices (ASEMD) (pp. 1–4)*. IEEE.

64. Sunwong, P., Higgins, J. S., Tsui, Y., Raine, M. J., & Hampshire, D. P. (2013). The critical current density of grain boundary channels in polycrystalline HTS and LTS superconductors in magnetic fields. *Superconductor Science and Technology, 26*(9), 095006.

65. Varlamov, A. A., Balestrino, G., Milani, E., & Livanov, D. V. (1999). The role of density of states fluctuations in the normal state properties of high T c superconductors. *Advances in Physics, 48*(6), 655–783.

66. Villamor, D. E. V., Ho, T., Al Rwahnih, M., Martin, R. R., & Tzanetakis, I. E. (2019). High throughput sequencing for plant virus detection and discovery. *Phytopathology, 109*(5), 716–725.

67. Wang, Y., Zheng, J., Zhu, Z., Zhang, M., & Yuan, W. (2019). Quench behavior of high-temperature superconductor (RE) Ba2Cu3Ox CORC cable. *Journal of Physics D: Applied Physics, 52*(34), 345303.

68. Werfel, F. N., Floegel-Delor, U., Riedel, T., Rothfeld, R., Wippich, D., & Goebel, B. (2006, June). Flywheel challenge: HTS magnetic bearing. In *Journal of Physics: Conference Series (Vol. 43, No. 1, p. 1007)*. IOP Publishing.

69. Wimbush, S. C., & Strickland, N. M. (2016). A public database of high-temperature superconductor critical current data. *IEEE Transactions on Applied Superconductivity, 27*(4), 1–5.

70. Wu, Z., Jin, J., Shen, B., Hao, L., Guo, Y., & Zhu, J. (2022). Fundamental Design and Modelling of the Superconducting Magnet for the High-Speed Maglev: Mechanics, Electromagnetics, and Loss Analysis during Instability. *Machines, 10*(2), 113.

71. Yamada, Y., Isojima, S., Masuda, T., Hayashi, K., Sato, K. I., & Hata, R. (2005, June). Recent development of high-temperature superconducting (HTS) cable in Sumitomo Electric Industries, Ltd. In *CIRED 2005–18th International Conference and Exhibition on Electricity Distribution (pp. 1–5)*. IET.

72. Zareapour, P., Hayat, A., Zhao, S. Y. F., Kreshchuk, M., Jain, A., Kwok, D. C., & Burch, K. S. (2012). Proximity-induced high-temperature superconductivity in the topological insulators Bi2Se3 and Bi2Te3. *Nature Communications, 3*(1), 1056.

73. Zhou, P., He, J., Huang, L., Yu, Z., Su, Z., Shi, X., & Zhou, J. (2020). Microfluidic high-throughput platforms for discovery of novel materials. *Nanomaterials, 10*(12), 2514.

74. Zhuravel, A. P., Sivakov, A. G., Turutanov, O. G., Omelyanchouk, A. N., Anlage, S. M., Lukashenko, A., & Abraimov, D. (2006). Laser scanning microscopy of HTS films and devices. *Low Temperature Physics, 32(*6), 592–607.

INDEX

W

X

Y

Z

Printed in the United States
by Baker & Taylor Publisher Services